Z-Theory and its Applications

Allan Zade

authorHOUSE®

AuthorHouse™
1663 Liberty Drive
Bloomington, IN 47403
www.authorhouse.com
Phone: 1-800-839-8640

First published by AuthorHouse 5/5/2011

ISBN: 978-1-4520-1893-5 (sc)

Library of Congress Control Number: 2011907271

Printed in the United States of America

Any people depicted in stock imagery provided by Thinkstock are models, and such images are being used for illustrative purposes only. Certain stock imagery © Thinkstock.

This book is printed on acid-free paper.

Z-Theory
and
Its Applications

by Allan Zade

Contents

Foreword

From its earliest time mankind has tried to understand and explain a lot of different events. Different times offered different explanations for the same events and curiosities. But some events continued to be unexplainable using any known reasoning.

Gradually science's system of exploration and explanation of events possessed the most power among all methods. Various civilizations founded different areas of knowledge and science, leading to modern times enjoying the well-developed fields of physics, chemistry, and mathematics, among others.

Step by step, new theories built a path through a mire of old mistakes, illuminating new ways and ideas, forming modern science's point of view. Extending its power, science possesses explanations for almost any event that mankind observes in nature or at laboratories. Scientists started to transfer their primary activity to laboratories, paying less attention to the surrounding world and believing that they were able to find an appropriate answer to any question.

Everything went well until mankind began to notice some types of phenomena that were they said were "impossible to explain" or "impossible to exist". Neither man nor science have yet any acceptable explanation for such phenomena because the events are in controversy with modern science's point of view and logic.

The causes and reasons of such phenomena lie far away from the horizon of modern science. To reach a higher point to where we can understand such phenomena, we need to go "through the horizon" that now restricts our knowledge and imagination. This book is dedicated to such a cruise in an aim to explain phenomena of some particular areas of physics and their consequences.

How to Read This Book

Unlike usual publications, this book contains a lot of cross-references. To simplify using all its information, I decided to use special number marks showing items' numbers. Each item is able to contain one or more paragraphs according to the type of information the item provides.

Each number contains the number of chapters, the number of captions (as part of chapter), and the number of the item itself. If an item contains smaller elements, they are mentioned by letters. Such a referencing system helps me to use references from any part of this work to an exact item. Each reference is enclosed in square brackets to simplify its searching.

For example, reference [1.2. 3] refers to chapter 1, part 2, item 3. Looking at this item, we can see the following: "Some time passes, and the first ship disappeared from human vision..." Reference to any smaller element looks like this: [2.3.8. a]. That means equation A that is shown at item 2.3.8.

Types of references can be mentioned as well as reference itself. Here are some types of possible references and the letters associated to them (in order of decreasing reference capacity): "because (b), see (s), remember (r)". If an item has the strongest reference to another, it can be marked by letter *b*, showing this item can be understood as a logical lead from other items.

If a cross-reference shows some relation to information mentioned at a different item, it can be marked by letter *s*. In cases of weak relation between items relation, it can be marked by letter *r*.

For example, reference [b.3.3.2] means, "That is correct **because** something mentioned at item 3.3.2 leads to this result (consequences)." Reference [s.3.3.2] means, "**See** item 3.3.2 to find more related information."

References inside the same item can be used omitting numbers. For example phrase "figure [A] shows..." means one should refer to the figure shown at same item in the phrase mentioned.

References to multiple items are usually mentioned by numbers of first and last item. For example, reference [b. 7.3.1–7.3.6] means, "That is correct **because of** something mentioned at items from 7.3.1 to 7.3.6."

Prolegomena

Chapter 1. Appearance of Problem

1.1 Boeing 727 Incident

1.1. 1. One of most famous and unusual incidents happened at the Miami Airport with a regular flight mentioned by Mr. Berlitz at his book *The Bermuda Triangle*. According to the book, the event contained the following steps.

1.1. 2. The airplane coming to the airport suddenly disappeared from the tracking radar screen and then reappeared ten minutes later.

1.1. 3. Subsequently the airplane landed without any incident.

1.1. 4. As it was found later, all onboard watches (as well as any other time indicators) were for ten minutes behind the airport clocks.

1.1. 5. In other words, the onboard watches did not register the time passing during the airplane's disappearance.

1.1. 6. Nobody onboard noticed anything unusual during the flight.

1.1. 7. Clear evidence:

   a. There was the aircraft.
   b. There was a flight in which the aircraft was involved.
   c. The airport was the flight's destination.
   d. The radar of the airport saw the flying aircraft.
   e. Event E occurred and existed at some particular time (it occupied some interval of time).
   f. The aircraft and event E existed at the same particular area of Earth's atmosphere at the same time interval.
   g. The radar of the airport saw no flying aircraft during event E.
   h. Ten minutes passed.
   i. Event E terminated.
   j. The radar of the airport saw the flying aircraft again.
   k. The aircraft successfully landed on the airport runway.
   l. All onboard watches of the aircraft indicated time ten minutes behind the time indicated by airport watches and clocks.
   m. Nobody onboard noticed anything strange.

1.2 Two-ship Incident

1.2. 1. The next very unusual incident happened in the Atlantic and involved two ships. The event is well described by Mr. Berlitz in the same work mentioned above. One of the ships had some problems with its engine and stopped moving.

1.2. 2. During the time the crew of the first ship was fixing their problems, the crew of the next ship had strange observations. First the ship became invisible by radar but was still visible by human observation.

1.2. 3. Some time passed, and the first ship disappeared from human vision. Then some time later the first ship reappeared again safe and sound. Onboard eyewitnesses of that ship noticed nothing unusual.

1.2. 4 Clear evidence:

a. There was a ship A.
b. There was a ship B.
c. There was event E.
d. The ship A and event E existed at the same particular area of Earth surface at the same time interval.
e. Human eyewitnesses from ship B saw ship A.
f. Radar of ship B saw no ship A.
g. Crew of ship B had seen nothing unusual.
h. Some time later the ships A and B changed their positions.
i. On changed position (and time), ship B became visible again for radars and visual observations.

1.3 Flight 19 Incident

1.3. 1. One of the best known Bermuda Triangle incidents concerns the loss of Flight 19, a squadron of five Navy bombers on a training flight out of Ft. Lauderdale on December 5, 1945. According to Mr. Berlitz, the flight consisted of expert pilots who, after reporting a number of odd visual effects and following the malfunction of onboard navigation instruments, simply disappeared.

1.3. 2. Furthermore, Mr. Berlitz claims that because the TBM Avenger bombers were built to float for long periods, they should have been found the next day considering the reported calm seas. However, not only were they never found, but a Navy search and rescue plane that went after them was also lost.

1.3. 3. No logical or physical explanation of that incident given can be recognized as an acceptable one.

1.3. 4. Clear evidence:

a. There were a number of aircrafts (group A)
b. One day all those aircrafts were involved in a flight.
c. The squadron commander noted the malfunction of the compasses of all aircrafts.
d. The squadron commander sent a message about malfunctioning instruments to the base by radio.
e. Radio transmission was successfully received by base.
f. In some area the pilots of those aircrafts had seen a number of odd visual effects.

g. The pilots were capable of sending information about those observations to the base.
h. The base was capable of receiving radio transmissions from the aircrafts.
i. One more aircraft (group B) was sent to the zone where the aircrafts (group A) were located.
j. There was additional radio contact between aircrafts (group A) and the base.
k. The aircrafts of group A disappeared.
l. One aircraft of group B disappeared.
m. No evidence of aircraft crash from group A or B was ever discovered.

Z-Theory

Chapter 2. Matter of Imagination

2.1 Addition Terminology

2.1. 1. You probably already noticed that all the events described in the previous chapter are more or less unusual. Modern people with scientific points of view rarely believe in miracles. As a result such incidences have never been analyzed by any mind that understands modern physics quite so well. The main reason for no investigation of the facts mentioned above was the strong position of scientists who denied even the possibility of such events happening.

2.1. 2. Nevertheless some number of strange events can be analyzed using well-known facts, descriptions of physical processes mentioned on scientific works, and books describing the surrounding world and basic physical principles. Using all those sources, we can obtain much more information from the description of any event and better understand it.

2.1. 3. Before we go further, I'd like to introduce some notions that help us describe events easier and understand them more clearly.

    a. Independent Bystander (IB) means a person or thing (device) that is present in our world but *not involved* with the event

    b. Independent Observer (IO) is a person or thing (device) that observes and *is involved* with the event.

    c. Independent Spectator (IS) means any combination of independent bystanders and/or independent observers.

To explain the difference between observations of the same object or event by IB and IO, we use the following hypothetical experiment.

2.1. 4. Figure A shows a diagram including one vertical and one horizontal stripe. The horizontal stripe has red color (showed on the figure by horizontal filing). The vertical stripe has blue color (showed on the figure by vertical filings).

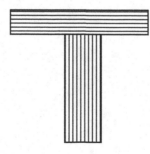

Fig. A.

There are three persons, A,B, and C. Two of them have glasses with color light filters exactly matching the colors of the figure stripes. Person A uses glasses with red light filter, person B uses glasses with blue light filter, and person C uses no glasses.

2.1. 5 Experiment begins. Persons A and B put on their glasses. Figure [2.1. 4. A] is shown to all three persons, and they are asked to describe what they see. As a result, person A declares one dark vertical stripe, person B declares of one dark horizontal stripe, and person C declares a two-colored cross.

2.1. 6. As you can now understand, there were differences in observations caused by previously occurring events (using glasses with color light filters). That simple event caused complex results to vision. The only person who was unaffected by these preceding events was person C. Only that person could see the unchanged picture and was able to correctly describe the image.

2.1. 7. According to [2.1. 5–2.1. 6] we have two independent observers (persons A and B) and one independent bystander (person C). The most important result of the [2.1. 5] experiment is this: two independent observers can see different images simultaneously according to the difference of their involved events (using different colors of glasses light filters).

2.1. 8. In the real world, a person usually doesn't know which role he or she plays. Is it the role of bystander or observer? Has she clear vision, or vision already changed by some strange event? What's more, the role of bystander can be easily changed to the role of observer when a person becomes involved in an event. That leads us to following question.

2.1. 9. Is it possible to separate declarations of observers and bystanders? In other words, where exists the observer and bystander? Certainly that is a very complex question. Generally nobody has a firm answer to it. But for the purposes of this work, it's possible to use the following consideration for the notion of an event.

2.1. 10. Event means any process that occurs in some particular place and can be noticed by the senses of someone or something.

2.1. 11. Anybody or anything that has *any sense* of the event is an Independent Observer.

2.1. 12. Anybody or anything that has *no sense* of the event is an Independent Bystander.

2.1. 13. According to [2.1. 11–2.1. 12], any event has some number of IO and IB, (including zero, at times). The difference between them depends on some particular zone that separates IO from IB. Usually that zone exists according to the sense range of a spectator. That consideration leads to the following definition.

2.1. 14. D-Radius (radius of detection) means the largest distance from which the spectator is able to have any sense of the happening event.

2.1. 15. An independent spectator (IS) that exists to the event closer than D-Radius becomes an independent observer (IO); otherwise it becomes an independent bystander (IB). That happens because senses of IO are affected by some changes produced by events in the surrounding world. Senses of IB are not affected by the same event. As a result IB is unable to detect the presence of an event.

2.2 Observable Facts of Air

2.2. 1. It's time to discuss the event described in [1.1]. At first it looks like a jamming of information without any possible way to analyze the event step by step. But if we use well-known facts from flight theory, we can see a lot of interesting conclusions.

Figure A put below shows the number of forces and their directions as applied to a flying aircraft.

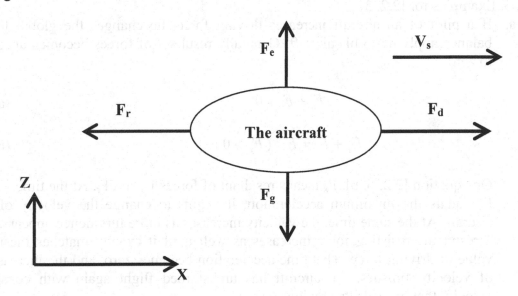

Fig. A

Vectors mentioned on figure [2.2. 1. A] have the following meaning (X and Z are axes of reference):

    a. F_d is vector of driving force
    b. F_e is vector of elevating force
    c. F_r is vector of resisting force
    d. F_g is vector of gravity force

2.2. 2. Vector V_s means the velocity direction of flying aircraft. Usually in undisturbed flight, vectors V_s and F_d coincide; vectors F_d and F_r lie on one straight line (DR-line) and have equal absolute value and opposite directions; vectors F_e and F_g lie on one straight line (EG-line) and have equal absolute value and opposite directions.

DR-line and EG-line are mutually perpendicular. DR-line and Earth surface below the aircraft are parallel. EG-line and Earth surface below the aircraft are mutually perpendicular.

2.2. 3. As long as $F_d+F_r=0$ and $F_e+F_g=0$, the resultant forces are equal to zero and an aircraft has zero acceleration to any direction (second law of Newton).

2.2. 4. In the case of [2.2. 3], an aircraft keeps constant speed and direction. As a result all onboard passengers (IO) have the sensation of constant flight (undisturbed flight).

2.2. 5. If any force mentioned on [2.2. 1. A] has any disturbance and changes its magnitude, and/or the direction of aircraft fight is immediately affected, according to new circumstances it will be changed until it reaches stabilization on a new state. Sometimes stabilization cannot ever be reached because of the force.

2.2. 6. Examples for [2.2. 5]:
   a. If a pilot of an aircraft increases driving force, he changes the global force balance applying to his aircraft. Generally resultant of forces becomes unequal to zero.

$$F_d + F_r \neq 0 \qquad \text{(a)}$$

$$F_d + F_r = F_x ; (F_x > 0) \qquad \text{(b)}$$

On equation [2.2. 6. b], F_x means resultant of forces F_d and F_r. At the time when F_x equals the maximum acceleration, it begins to change the velocity of an aircraft. At the same time, the velocity increases as more turbulence appears. As a result the resisting force increases as well until it exactly matches the new value of driving force. That time acceleration becomes zero, and the increasing of velocity finishes. An aircraft has undisturbed flight again with constant velocity that exceeds its previous one.
   b. If a pilot of an aircraft changes the elevating force of any wing, it causes moments of force appearance. The force of one wing exceeds the force of the other one. As a result an aircraft begins to rotate around the longitudinal axis (X axis on Fig. [2.2. 1. A]). That rotation never stops until a pilot readjusts the elevating force of the appropriate wing and makes elevating forces of both wings equal to each other.

2.2. 7. Consideration of [2.2. 5–2.2. 6] is applicable for changing of any force mentioned at [2.2. 1–2.2. 5].

2.2. 8. According to [2.2. 1–2.2. 7] we can see following result: If any force applied to an aircraft on undisturbed flight has any changes, it causes a change on the flight's state. As a result passengers and crew of an aircraft can feel some changes. Those changes can be detected as rising or falling velocity (horizontal or vertical), direction of flight, or angling of an aircraft (angular momentum).

2.2. 9. Generally any significant physical impact on an aircraft that can be reduced to single force (or described as single force) produces changes on aircraft trajectory. Such changes would be noticed by onboard humans. If they noticed no changes on the aircraft's flight at any exact point of time, there is no additional force that is applied to an aircraft at that time.

2.2. 10. **Summary**. If no onboard person noticed any changes on aircraft trajectory or velocity during the event mentioned at [1.1], no additional force was applied to the aircraft at that time.

2.2. 11. There are some examples of physical processes inside and around the aircraft that were unchanged before, during, and after event mentioned at [1.1]. These are some considerations according to [2.2. 10]:

a. Air flow around the aircraft
b. Power supply of turbines
c. Temperature of burning fuel in turbines
d. Volume of fuel provided to each turbine
e. Power supply of fuel pumps
f. Power supply of electric generators provided onboard electricity
g. Power supply of onboard computers
h. Operation of onboard computers
i. Operation of onboard navigation system

It looks like the event mentioned at [1.1] changed nothing onboard and around the aircraft.

2.3 Types of Forces

2.3. 1. Figure [2.2. 1.A] shows the combination of four forces F_d, F_r, F_e, and F_g. Those forces described are above, and now I'd like to mention their origins.

2.3. 2. Driving force (F_d) exists accordingly with the power of the aircraft engine. Generally it can be any type of engine used in aviation (jet, propeller, etc.). Each type of engine provides an aircraft with necessary propelling power and produces enough driving force. Obviously, that force has a handmade origin and is controlled by the aircraft crew.

2.3. 3. Elevating force (F_e) exists accordingly with the shape and type of wings, and the speed of air flow around them. Different types of wings produce different values of elevating force at the same speed. Additionally, pilots can change elevating force to some limits. Usually they use such opportunity to change the aircraft from one mode of flight to another one (takeoff, landing, flight mode, etc.). Obviously, that force has a handmade origin and is controlled by the aircraft crew.

2.3. 4. Resisting force (F_r) exists accordingly with the shape, size, and proportion of an aircraft. Generally that force is produced by air flow around the aircraft and exists as force proportional to velocity of an aircraft. Usually pilots can't change that force. Moreover, they do need not to change resisting force because its presence is derived from aircraft motion.

There is only one case where that force can be changed directly: the aircraft's run after landing. High-speed vehicles can be stopped on an acceptable length on runway using only special air breaks, such as reversible turbines and brake parachutes. Such breaks are used only after landing and never on regular flight.

Nevertheless resisting force has a handmade origin because of the aircraft's construction.

2.3. 5. Gravity force (F_g) exists by itself. The source of that force is the gravity field surrounding the Earth as well as any other celestial body (Newton's law of gravity). That field produces special forces according to the law of gravity invented by Isaac Newton. The following equation represents that law mathematically.

$$F_g = G \cdot \frac{m_1 \cdot m_2}{r^2}$$
(a)

Newton's law of gravitation is statement that any particle of matter in the universe attracts any other with a force varying directly as the product of the masses and inversely as the square of the distance between them. In symbols, the magnitude of the attractive force F is equal to G (the gravitational constant, a number the size of which depends on the system of units used and which is a universal constant) multiplied by the product of the masses (m_1 and m_2) and divided by the square of the distance R: $F = G(m_1 m_2) / R^2$. Isaac Newton put forward the law in 1687 and used it to explain the observed motions of the planets and their moons, which had been reduced to mathematical form by Johannes Kepler early in the 17th century.[1]

2.3. 6. Obviously, the origin of gravity force is not a handmade one. Even the masses of the aircraft's parts are given to them by nature. Engineers can use different construction elements with different masses, but it's impossible to change the mass of any given aircraft element during regular flight.

2.3. 7. As you can see, this is the only force applied to an aircraft on undisturbed flight regardless of human wishes. The origin of that force is our planet, which holds us tightly to Earth's surface. The same force drags an aircraft to Earth's surface (as well as any other object) at a constant rate regardless of flight.

2.3. 8. The only difference between an aircraft flying and staying on the runway is the volume of elevating force. If the forces of elevation and gravity are equally to each other, the vehicle is able to fly without any disturbance as long as the following equations persist:

$$F_d + F_r = 0$$
(a)

---

[1] "Newton's law of gravitation." Encyclopaedia Britannica. Encyclopaedia Britannica 2008 Deluxe Edition. Chicago: Encyclopaedia Britannica, 2008.

$$F_e + F_g = 0 \qquad\qquad (b)$$

Where F_d is driving force, F_r is resisting force, F_e is elevating force, and F_g is gravity force.

2.4 Force Diagram

2.4. 1. Let's look more closely at figure [2.2. 1.A]. If we use the diagram to describe forces applied to other type of vehicles, we see the possibility to use the mentioned number of forces on the description of any vehicle's state. The following paragraphs show a modification of figure [2.2. 1.A] according to other types of vehicles.

2.4. 2. The following figure shows number of forces applied to any watercraft. The only difference between Fig. [2.2. 1. A] and Fig. [2.4. 2. A] is this: a watercraft uses buoyancy force instead of elevating one to compensate gravity force.

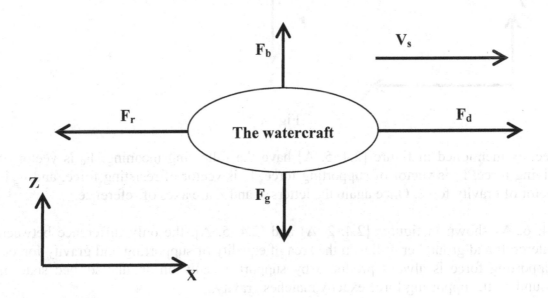

Fig. A

Vectors mentioned on figure [2.4. 2. A] have the following meaning: F_d is vector of driving force, F_a is vector of buoyancy force, F_r is vector of resisting force, and F_g is vector of gravity force. The letters X and Z are axes of reference.

2.4. 3. The only difference between watercraft and submarines is the area of equality of buoyancy and gravity forces. A watercraft is able to keep balance with those forces only on the water's surface. Any vertical disturbance increases or decreases the buoyancy force and changes the balance between that force and gravity. As a result a watercraft rises or falls according to a new level of surrounding water until it possess a new exact balance.

2.4. 4. A submarine is able to keep balance between vertical forces anywhere inside water. As a result it can stay at any acceptable depth. If the submarine submerges deeper than the accepted level, it can be crushed by the high pressure of ocean water. That process lies outside of this work's purpose.

2.4. 5. The following figure shows the number of forces applied to any craft moving on a solid surface of the Earth.

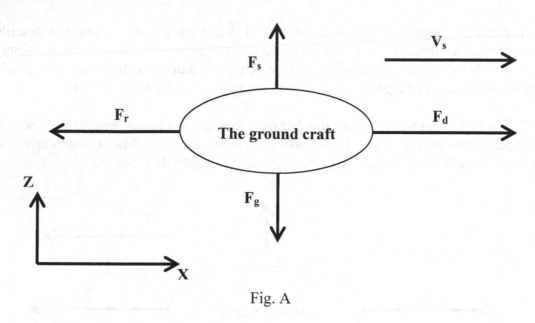

Fig. A

Vectors mentioned in figure [2.4. 5. A] have the following meaning: F_d is vector of driving force, F_a is vector of supporting force, F_r is vector of resisting force, and F_g is vector of gravity force. Once again the letters X and Z are axes of reference.

2.4. 6. As shown in figures [2.4. 2. A] and [2.4. 5. A], the only difference between watercraft and ground craft lies in the area of equality of supporting and gravity forces. Supporting force is always produced by support. usually on an undisturbed state of ground craft, supporting force exactly matches gravity.

2.4. 7. **Summary**. As you can see from part [2.4], there is an equal number of forces applied to any moving craft regardless of its nature. Each craft has two pairs of forces that mutually compensate each other. As a result moving craft can keep undisturbed motion as long as no additional force is applied to it.

The nature of the three forces (driving, supporting, and resisting) mentioned above depends on the type of a craft. But there is only force with its own nature that exists regardless of craft type: the force of gravity. It depends only on the mass of a craft and exists as long as the craft itself.

2.4. 8. Generally gravity is involved in any process applied to any object. All types of crafts mentioned previously are examples of more or less massive objects existing in the gravity field of the Earth. That leads us to following conclusions:

a. Any event must leave some trace on the gravity field of surrounding objects.
b. The gravity field must be involved in any event that happens to any object.

Chapter 3. Matter of Gravity
3.1 Newton's Gravity

3.1. 1. According to the modern view on gravity fields, we have a basic equation describing the gravity force between two objects mentioned at [2.3.5. A]. Gravity itself can be described as the following.

3.1. 2 "Gravitation in mechanics, the universal force of attraction acting between all matter. It is by far the weakest known force in nature and thus plays no role in determining the internal properties of everyday matter. On the other hand, through its long reach and universal action, it controls the trajectories of bodies in the solar system and elsewhere in the universe and the structures and evolution of stars, galaxies, and the whole cosmos. On Earth all bodies have a weight, or downward force of gravity, proportional to their mass, which the Earth's mass exerts on them. Gravity is measured by the acceleration that it gives to freely falling objects. At the Earth's surface the acceleration of gravity is about 9.8 metres (32 feet) per second per second. Thus, for every second an object is in free fall, its speed increases by about 9.8 metres per second. At the surface of the Moon the acceleration of a freely falling body is about 1.6 metres per second per second."[1]

3.2 Strength of Gravity Force

3.2. 1. Equation [2.3.5. A] gives us the value of gravity force between two objects. Obviously, that equation can be used to calculate the exact force but not the gravity field itself. What if we'd like to describe the gravity field instead of the gravity force? It's quite possible if we use the known method of field description, the reduced force.

3.2. 2. Usually reduced force plays the role of a universal characteristic of any field. It describes how much force is produced by observing the field on a single object containing one piece of analyzing characteristic. "The value of the electric field at a point in space, for example, equals the force that would be exerted on a unit charge at that position in space."[2] That measure gives information of electric field force characteristics. "The strength of an electric field E at any point may be defined as the electric force F exerted per unit positive electric charge q at that point, or simply $E = F / q$. If the second, or test, charge is twice as great, the resultant force is doubled; but their quotient, the measure of the electric field E, remains the same at any given point."[3] The following equation shows the mathematical description for electric field force:

---

[1] "Gravitation." Encyclopaedia Britannica. Encyclopaedia Britannica 2008 Deluxe Edition. Chicago: Encyclopaedia Britannica, 2008.

[2] "Electromagnetism." Encyclopaedia Britannica. Encyclopaedia Britannica 2008 Deluxe Edition. Chicago: Encyclopaedia Britannica, 2008.

[3] "Electric field." Encyclopaedia Britannica. Encyclopaedia Britannica 2008 Deluxe Edition. Chicago: Encyclopaedia Britannica, 2008.

$$F = \frac{q_1 \cdot q_2 \cdot r}{4 \cdot \pi \cdot \varepsilon_0 \cdot r^3} \qquad\qquad (a)$$

"The ... characters F and r are vectors, F being the force which a point charge q_1 exerts on another point charge q_2. The combination r / r_3 is a vector in the direction of r, the line joining q_1 to q_2, with magnitude $1 / r_2$ as required by the inverse square law. When r is rendered in lightface, it means simply the magnitude of the vector r, without direction. The combination $4p\varepsilon_0$ is a constant whose value is irrelevant to the present discussion. The combination $q_1 r / 4p\varepsilon_0 r_3$ is called the electric field strength due to q_1 at a distance r from q_1 and is designated by E; it is clearly a vector parallel to r. At every point in space E takes a different value, determined by r, and the complete specification of E(r) – that is, the magnitude and direction of E at every point r – defines the electric field."[2]

If we need to calculate only the magnitude of force mentioned above, we can use the following equation:

$$F = \frac{q_1 \cdot q_2}{4 \cdot \pi \cdot \varepsilon_0 \cdot r^2} \qquad\qquad (b)$$

That equation shows the value of force between two charges q_1 and q_2 separated by distance r.

It's possible to produce force characteristics for gravity field using the same method. That leads us to following statement:

3.2. 3. Strength of gravity force (SGF) must be equal to the magnitude of force produced by the mass of an object on the unit mass positioned at a given distance from the object.

3.2. 4. To determine that force we need to calculate the interaction between an object with mass M and an object with unit mass. According to equation [2.3.5. A], we have the following equation:

$$S_g = G \cdot \frac{m_1}{r^2} \qquad\qquad (a)$$
$$m_2 = 1 \qquad\qquad (b)$$
$$m_1 \gg m_2 \qquad\qquad (c)$$

[1] "Physical science, principles of." Encyclopaedia Britannica. Encyclopaedia Britannica 2008 Deluxe Edition. Chicago: Encyclopaedia Britannica, 2008.

[2] "Physical science, principles of." Encyclopaedia Britannica. Encyclopaedia Britannica 2008 Deluxe Edition. Chicago: Encyclopaedia Britannica, 2008

In the equations mentioned above, S_g is strength of gravity force, M_1 is mass of an object, R is the separation between the object and analyzing point of space, and G is the universal gravitational constant.

Equation [A] helps us to formulate following statement.

3.2. 5. SGF at any point of space surrounding the object is proportional to the mass of the object and is inversely proportional to the square of the separation between the object and the analyzing point of space.

3.2. 6. To describe a pair of objects with different masses according to equation [3.2. 4. C], I use the following terminology:

a. The body is used to mention a very massive entity.
b. The object is used to mention an entity with very low mass comparative to the body.

3.2. 7. These examples show a number of various pairs of body and object:

a. A planet comparative to a craft
b. A star comparative to a planet
c. A galaxy comparative to a star

3.2. 8. Equation [3.2. 4. A] leads us to following important conclusion:

a. To calculate gravity force between the body and the object, we need to multiply the SGF of any exact point where the object exists to mass of the object.
b. Any point of space that lies equal distance from the body has equal SGF.
c. SGF rises continuously from any far point to any point positioned closer to the body.

3.2. 9. The next figure shows aspects [3.2. 8. b] and [3.2. 8. c] graphically.

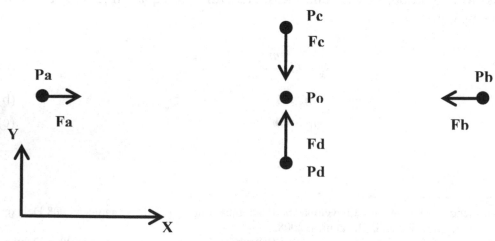

Fig. A

There are few points on figure [A]. Point (P_o) is the location of the body. Other points (P_a, P_b, P_c, P_d) are examples of analyzing points of gravity field. On the figure, OB = OA and OC = OD. As a result SGF on points mentioned above are SGF(a) = SGF(b) and SGF(c) = SGF(d). Other gravitational forces affect objects with unit mass mentioned on the figure by vectors F_a, F_b, F_c, and F_d. According to the distance between the body and analyzing points, $F_a = F_d$, $F_c = F_d$, $F_a < F_c$, and SGF $P_a = P_b$, $P_c = P_d$.

3.2. 10. Usually in figures the number of points around a body with an equal level of gravitational force and SGF are showed by circles surrounding the central body. As a result figure [3.2. 9.A] can be transformed to figure [3.2. 10. A].

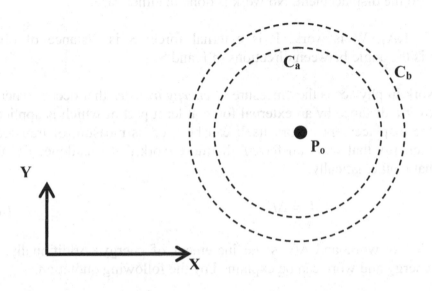

Fig. A

On the figure circle, A (C_a) means the number of points with equal SGF with value A. Circle B (C_b) means the number of points with different SGF with value B. SGF on circle A is higher than SGF of circle B. C_a and C_b are concentric circles with centers that coincide with the body\- producing gravitational field.

3.3 Work and Energy in Gravity Field

3.3. 1. According to the modern point of view on work produced by any force, there is an equation describing the relation between work and force:

$$W = F \cdot S \cdot \cos(\varphi) \tag{a}$$

Work in physics, measure of energy transfer that occurs when an object is moved over a distance by an external force at least part of which is applied in the direction of the displacement. If the force is constant, work may be computed by multiplying the length of the path by the component of the force

25

acting along the path. Work done on a body is accomplished not only by a displacement of the body as a whole from one place to another but also, for example, by compressing a gas, by rotating a shaft, and even by causing invisible motions of the particles within a body by an external magnetic force.

No work, as understood in this context, is done unless the object is displaced in some way and there is a component of the force along the path over which the object is moved. Holding a heavy object stationary does not transfer energy to it, because there is no displacement. Holding the end of a rope on which a heavy object is being swung around at constant speed in a circle does not transfer energy to the object, because the force is toward the centre of the circle at a right angle to the displacement. No work is done in either case.[1]

On equation [3.3. 1.A], W is work, F is external force, S is distance of object movement, and φ is the angle between directions of F and S.

3.3. 2. Because work in physics is the "measure of *energy transfer* that occurs when an object is moved over a distance by an external force at least part of which is applied in the direction of the displacement", work itself can be used as measure of transferred energy. The more energy that was transferred, the more work that was done. The next equation shows that mathematically.

$$W = \Delta E \qquad\qquad\qquad (a)$$

In the equation 'W' is work and ΔE is the increment of energy. Additionally, the relation between energy and work can be explained by the following quotation.

Energy is usually and most simply defined as the equivalent of or capacity for doing work. The word itself is derived from the Greek energeia: en, "in"; ergon, "work." Energy can either be associated with a material body, as in a coiled spring or a moving object, or it can be independent of matter, as light and other electromagnetic radiation traversing a vacuum. The energy in a system may be only partly available for use. The dimensions of energy are those of work, which, in classical mechanics, is defined formally as the product of mass (m) and the square of the ratio of length (l) to time (t): ml^2 / t^2. This means that the greater the mass or the distance through which it is moved or the less the time taken to move the mass, the greater will be the work done, or the greater the energy expended.[2]

Keeping that in mind, equation [3.3. 1.A] can be rewritten as follows:

---

[1] "Work." Encyclopaedia Britannica. Encyclopaedia Britannica 2008 Deluxe Edition. Chicago: Encyclopaedia Britannica, 2008.
[2] "Energy conversion." Encyclopaedia Britannica. Encyclopaedia Britannica 2008 Deluxe Edition. Chicago: Encyclopaedia Britannica, 2008.

$$\Delta E = F \cdot S \cdot \cos(\varphi) \qquad \text{(b)}$$

where ΔE is increment of energy, F is external force, S is distance of object movement, and φ is angle between directions of F and S

3.3. 3. Equation [3.3. 2.B] shows dependencies between increment of energy, external force, and parameter $\cos(\varphi)$ (cosines of angle between directions of force and motion). As we can see, energy keeps constant ($\Delta E = 0$) in the following cases:

    a.   The external force is equal to zero (F = 0)
    b.   The distance of object movement is equal to zero (S = 0)
    c.   The $\cos(\varphi)$ is equal to zero

Each time when one of the cases mentioned above has occurred, a moving object changes no energy. As a result, such a condition of any object can exist as long as no change appears in the world surrounding the object. Otherwise the object changes energy (transforms energy from one form to other one) and possess a limit of such state because any state that has changing energy stops to exist as soon as the object consumes all energy possible in the transformation.

3.3. 4. In a real world state, [3.3. 3.A] has no chance to exist because each point of the universe contains a gravity field. So an object located anywhere in universe has external forces applied to it because of gravity.

3.3. 5. State [3.3. 3.B] exists only when we have a motionless object. For example, a rock lying on the Earth's surface keeps motionlessness for any given period of time. That happens because the external force (gravity force in that case) exists, but the object involved is not in motion by that force.

3.3. 6. State [3.3. 3.C] means a special type of motion: there an object always moves instantly perpendicular to the external force. That happens when an object goes around the central body that produces the gravity field. If the trajectory of an object lies on an exact circle with the center coinciding to the central object, then direction of motion and direction of external force are mutually perpendicular.

3.3. 7. The following figure shows state [3.3. 3.C] graphically.

On figure [3.3.7.A], B is central body, O is moving object, V is vector of velocity of object O, F_g is vector of gravitational attraction applied to the object by central body, C_b is circle with center coincide to point B (central body).

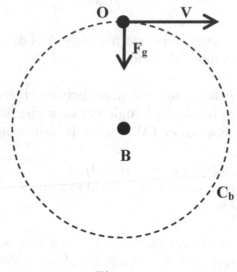

Fig. A

3.3. 8. Therefore, the energy of an object in a gravity field keeps constant as long as the object keeps its unchanging distance from the central body. As shown before, that happens because the gravity field is unable to produce work in case an object moves on a perpendicular vector to gravity force [b. 3.3. 2; 3.3. 6].

3.3. 9. As a result any location of an object around the central body with the same distance between the object and the body has no difference in energy of an object. In the real world the number of points with equal potential energy (of gravity field) forms a sphere around the central body.

3.4 Correlation between Strength and Potential Energy

3.4. 1. On one hand (as shown on [3.2. 10]), the SGF of any point located on any given distance from central body equals exactly the same value.

On the other hand (as shown on [3.3. 9]), the energy of an object positioned on any given distance from central body exactly equals the same value.

Both values are unequal to each other, but each value keeps constant as long as the object keeps constant distance from central body. Therefore, we can use constant measure of SGF value to have a measure of constant energy of an object.

3.4. 2. Two or more points in space around the central body with equal value of SGF relate to equal value of potential energy of any object with the same value of mass located on those points.

3.4. 3. The following figure shows statement [3.4. 2] graphically.

28

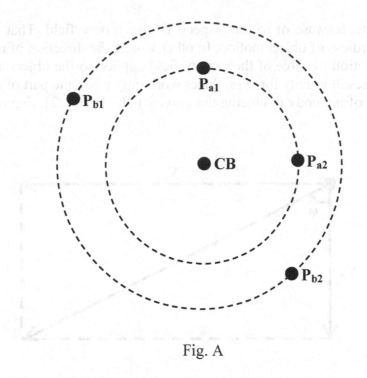

Fig. A

3.4. 4. Figure [3.4. 3.A] shows two concentric circles (A and B) around the central body (CB). Centers of each circle coincide to the center of central body. According to [3.4. 2], SGF of any point of the same circle equals to same value. As a result the next equations are applicable for the case shown in figure [3.4. 3.A].

$$S_g(P_{a1}) = S_g(P_{a2}) \tag{a}$$
$$S_g(P_{b1}) = S_g(P_{b2}) \tag{b}$$

In the equations, S_g is SGF on some particular point of space, P_{an} is any point of circle A, and P_{bn} is any point of circle B.

3.5 Motion in a Gravity Field

3.5. 1. This topic discusses closer motion of an object in the gravity field of a body. To go further we must remember the concept of conservative force.

3.5. 2. "Conservative force in physics, any force, such as the gravitational force between the Earth and another mass, whose work is determined only by the final displacement of the object acted upon. The total work done by a conservative force is independent of the path resulting in a given displacement and is equal to zero when the path is a closed loop."[1]

---

[1] "Conservative force." Encyclopaedia Britannica. Encyclopaedia Britannica 2008 Deluxe Edition. Chicago: Encyclopaedia Britannica, 2008.

3.5. 3. That happens because of natural aspects of the gravity field. That field has its own direction regardless of object motion. In other words, the direction of motion of an object and the direction of force of the gravity field applied to the object are mutually independent. As a result gravity force produces work only when the part of motion goes along the direction of the body producing the gravity field. [r. 3.3. 2]. Figure [A] shows that graphically.

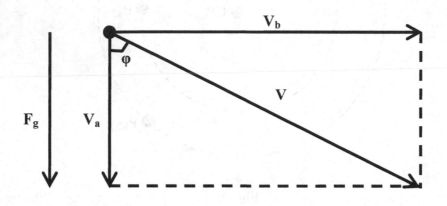

Fig. A

In the figure :
- F_g is vector of gravity force
- V is vector of velocity of an object
- V_a is component vector of velocity
- V_b is component vector of velocity
- ϕ is angle between direction of velocity of an object (V) and gravity force F_g
- Vectors V_a and V_b are mutually perpendicular
- Vectors V_a and F_g are parallel

3.5. 4. Because the magnitude of vector of velocity means displacement of a body per unit time, figure [3.5. 3.A] can be redrawn like this:

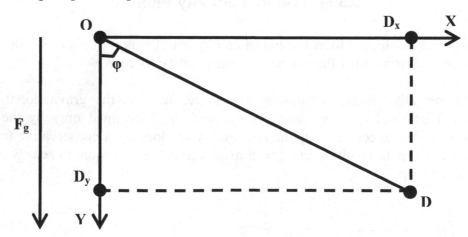

Fig. A

30

Where:

- F_g is vector of gravity force
- X and Y are axis of reference
- X and Y are mutually perpendicular
- Vectors Y and F_g are parallel
- O is beginning point of object position
- D is end point of object displacement
- D_x is projection of point D on X axe
- D_y is projection of point D on Y axe
- OD_x is projection of displacement OD on X axe
- OD_y is projection of displacement OD on Y axe
- φ is angle between displacement (OD) of an object and gravity force F_g
- φ is also angle between displacement (OD) of an object and OD_y

3.5. 5. Therefore, the work of the gravity field can be calculated as the sum of work produced by displacement along two mutually perpendicular axis (X and Y on fig. [3.5. 4.A]). Displacement of an object along X axis produces no work because the X axis is perpendicular to F_g [r. 3.3. 2. B]. Furthermore, displacement along Y axis produces all amount of work by gravity force. That happens because projection of displacement coincides with vector of gravity force. With that case equation, [3.3. 2.B] becomes equal to its maximal value ($\cos(\varphi) = 1$).

3.5. 6. Thus displacement of an object along the X axis (perpendicular to F_g) changes no energy of an object. So any object can be moved for any distance from its initial position using only displacement in the gravity field, changing no potential energy between two different points with equal value of S_g.

3.5. 7. Energy of an object can be only changed in cases where projection of displacement on direction of gravity force are unequal to zero. In the case of positive projection, gravity force produces positive work and increases the energy of an object [b. 3.3. 2. B; when $\varphi < 90$ degree]. In the case of negative projection, gravity force produces negative work and decreases energy of an object [b. 3.3. 2.B; when $\varphi > 90$ degree].

In the case of a free-falling object, that law leads to changing vertical component of moving object velocity, transforming potential energy of gravity field to kinetic energy of the falling object and vice versa (if the vertical component of velocity has an opposite direction relative to the direction of gravity force).

3.5. 8. As a result any motion of an object, where it has the same displacement with positive projection and negative projection with equal magnitude, produces work equal to zero. Hence, such motion changes no energy of an object.

3.5. 9. Therefore, the relationship between trajectory and changing of energy of an object by motion in gravity field lies only in the area of projection of full length of motion on the direction of gravity force. In the case of curved trajectory, it can be split to the number of elementary pieces of little trajectories that can be projected to the direction of gravity force independently. If the sum of such projections equals zero, the force of gravity field produces no energy by moving an object for any trajectory.

3.5. 10. Using [3.5. 9], statement [3.4. 2] can be rephrased following way.

3.5. 11. Two or more points in space around a central body with equal value of SGF relative to equal value of the full energy of any object *exists* on those points (in cases of free-falling objects).

3.5. 12. Statement [3.5. 11] introduces a principle of equivalence of any point with the same value of SGF. In other words, if an object is displaced from any point with exact value of SGF to any other point with same value of SGF, the full energy of that object remains constant regardless of the trajectory it moves along.

3.5. 13. The same statement [3.5. 11] can be written mathematically:

$$E_g(P_{a1}) = E_g(P_{a2}) \tag{a}$$
$$E_k(P_{a1}) = E_k(P_{a2}) \tag{b}$$
$$\Delta E_g = E_g(P_{a1}) - E_g(P_{a2}) = 0 \tag{c}$$
$$\Delta E_k = E_k(P_{a1}) - E_k(P_{a2}) = 0 \tag{d}$$
$$\Delta E = \Delta E_g + \Delta E_k = 0 \tag{e}$$

In the equations, P_{a1} and P_{a2} are two independent points of space surrounding the central body with equal value of SGF (equally elevated points). (P_{a1} and P_{a2} have same the meaning in equation [3.5. 13.A] and figure [3.4. 3.A].)
ΔE_g is the changing of potential energy of an object produced by gravity force.
ΔE_k is the changing of kinetic energy of a moving object produced by gravity force.

3.5. 14. <u>Conclusion</u>. Gravity field helps an object to be moved from one point of space to another with equal value of SGF without changing the energy of an object and the gravity field itself, regardless of the object's full trajectory length.

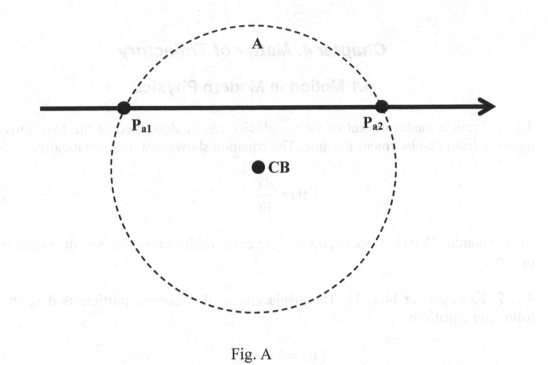

Fig. A

Figure [A] shows that graphically. P_{a1} and P_{a2} are two points of equal SGF. If an object moves in a straight line as shown in the figure (or on any other trajectory [r. 3.5. 11]), it has equal energy on points of equal SGF mentioned on figure as P_{a1} and P_{a2}.

Chapter 4. Matter of Trajectory

4.1 Motion in Modern Physics

4.1. 1. From a modern point of view, velocity can be described as the first derivative quantity from displacement for time. The equation shows that mathematically.

$$V(t) = \frac{dX}{dt} \qquad \text{(a)}$$

In the equation, V(t) is velocity, dX is differential of displacement, and dt is differential of time.

4.1. 2. Example for [4.1. 1]: The displacement of a moving particle is described by following equation:

$$X(t) = 3 \cdot t + 7 \qquad \text{(a)}$$

and the velocity of the particle is described by following one:

$$V(t) = \frac{d(3 \cdot t + 7)}{dt} = 3 \qquad \text{(b)}$$

Equation [4.1. 2.B] shows that there is no relation between velocity magnitude and time. That means the particle always moves with constant speed.

4.1. 3. As you can see from [4.1. 1.A], there is a strong rule of the calculation of velocity. To get a result from that calculation, we need to use differentiable displacement and time. Therefore, we have the following conclusion.

4.1. 4. Notion of displacement and all its derivative notions (velocity, acceleration, etc.) are only applicable to a particle that has *real world trajectory*.

4.1. 5. Real World Trajectory (RWT) is any trajectory where the location of a particle has only one exact point for any given moment of time.[1]

4.2 Motion in a Gravity Field

4.2. 1. A closer look at figure [3.5. 14.A] leads us to following important question.

---

[1] For example: calculation of equation [4.1. 2.A] gives us only value X(t) for any value of t.

4.2. 2. If the energy of a moving particle passing two different points of space keeps a constant value regardless of the trajectory particle used, is trajectory itself necessary for the description of particle displacement?

4.2. 3. As shown in [3.5. 13. E], nothing changed that way, including the energy of the moving particle. In other words, changing energy in the system equals zero in that case. This can be shown mathematically.

$$\Delta E = 0 \qquad \text{(a)}$$

4.2. 4. As a result, figure [3.5. 14.A] can be redrawn the following way:

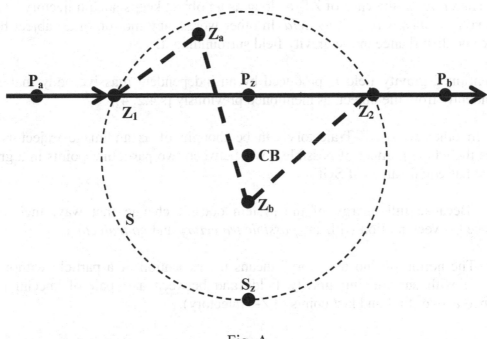

Fig. A

Where:
- Z_1 and Z_2 are points with equal SGF
- P_a, Z_1, Z_2, and P_b are points of RWT of a particle
- CB is central body
- S is number of points with equal SGF around the central body
- S_z is a particular point of S

4.2. 5. According to the considerations mentioned above [4.2. 3], there is no difference in the energy of a particle that moves on trajectory P_a, Z_1, P_z, Z_2, P_b (RWT) and P_a, (Z_1-Z_2), P_b. The difference between those two trajectories is the motion of a particle on its first trajectory (RWT), which means observable interaction between a particle and a gravity field on each point of its trajectory.

35

4.2. 6. In other cases, when part of (RWT) contains some displacement, then a particle has no observable interaction with a gravity field (Z_1-Z_2), but a particle location on its first and last points (moving between those points) of that displacement keeps zero difference of its whole energy, and we have special parts of displacement that changes no energy in the whole system. Here and further I refer such part of trajectory as Z-Trajectory (or ZT).

4.2. 7. Z-Trajectory is drawn on figure [4.2. 4.A] as trajectory of Z_1, Z_a, Z_b, Z_2. That is, image trajectory connects two points of the gravity field with equal magnitude of SGF by means of a special number of points that has no interaction with the surrounding gravity field.

4.2. 8. Therefore, in the case of ZT, as long as an object keeps such trajectory, *it can't be observed by means of gravity field.* In other words, any motion of an abject by ZT produces no disturbance in the gravity field surrounding it.

The mentioned gravity field is produced by an independent massive body that exists independently from the object, as mentioned previously [s.4.2. 4].

4.2. 9. In other words, Z-Trajectory can be thought of as an image trajectory that replaces the whole number of possible RWT between two particular points in a gravity field that has equal values of SGF.

4.2. 10. Because full energy of the system doesn't change that way, there is no difference between motion with any *possible trajectory* and *no trajectory*.

4.2. 11. The notion of "no trajectory" means the relocation of a particle without any interaction with surrounding gravity fields and between any pair of special points mentioned above (first and last points of Z-Trajectory).

Chapter 5. Matter of Conservative Fields

5.1. 1. Parts 3 and 4 are based on a key feature of gravity field its conservativeness. As mentioned in [3.5. 2], any conservative field has the same type of force characteristics. The only difference between them is the value of force (magnitude of force) that is produced on the appropriate unit. "The inverse square laws of gravitation and electrostatics are examples of central forces where the force exerted by one particle on another is along the line joining them and is also independent of direction. Whatever the variation of force with distance, a central force can always be represented by a potential; forces for which a potential can be found are called conservative."[1]

5.1. 2. Therefore, all consideration that can be correctly applied to a gravity field in the area of force value and energy according to mass unit can be correctly applied to the electric field in an area of force value and energy according to the unit charge.

5.1. 3. As mentioned in [3.2. 2], the electric field has a special measure of force producing to the unit charge that is placed in any electric field. That measure has the name of electric field strength.[2] To unify symbols using this work, the following equation describes equality between E and S_e.

$$S_e = E \qquad \text{(a)}$$

S_e is strength of electric field, and E is strength of electric field as mentioned on [3.2. 2].

5.1. 4. In the case of presence of two types of conservative forces simultaneously at any given point of space, as well as a particle that accepts force from each field (an electrically charged particle), we have the following equation of full strength of the surrounding field for a particle.

$$S = S_g + S_e \qquad \text{(a)}$$

Where:
- S is full strength of fields at given point of space
- S_g is strength of gravity field at given point of space
- S_e is strength of electric field at given point of space

5.1. 5. Theoretically in the case of the presence of more types of simultaneous conservative forces, equation [5.1. 4.A] can be rewritten as following:

---

[1] "Physical science, principles of." Encyclopaedia Britannica. Encyclopaedia Britannica 2008 Deluxe Edition. Chicago: Encyclopaedia Britannica, 2008.
[2] "Electric field." Encyclopaedia Britannica. Encyclopaedia Britannica 2008 Deluxe Edition. Chicago: Encyclopaedia Britannica, 2008.

$$S = S_g + S_e + S_c \qquad \text{(a)}$$

Where S_c is sum of strength of other type of any number of conservative forces.

5.1. 6 In the case of presence, only field strength of other fields becomes equal to zero, and equation [5.1. 5.A] turns to three particular cases.

$$S = S_g; \; S_e = S_c = 0 \qquad \text{(a)}$$
$$S = S_e; \; S_g = S_c = 0 \qquad \text{(b)}$$
$$S = S_c; \; S_g = S_e = 0 \qquad \text{(c)}$$

Usually case (a) is the most probable one because as far as modern physics is concerned, each part of space contains some value of a gravity field. That is the main reason why I mentioned only gravity field and related facts in previous parts of this work.

Cases (b) and (c) look more theoretical and describe the strength of fields without a gravity field involved. But in the presence of a gravity field absence they become useless.

5.1. 7. Here and further on, I refer to S as *full strength of conservative fields*, whose value can be calculated by equation [5.1. 5. A]. Hence, [4.2. 7] becomes the following.

5.1. 8. Z-Trajectory is the image trajectory that connects two points of space with equal value of full strength of conservative fields by means of a special number of points that have no interaction with surrounding conservative fields that are produced by independent bodies.

5.1. 9. Therefore, in the case of ZT, as long as an object keeps such trajectory, it can't be observed by means of any conservative field. In other words, any motion of an object by ZT produces no disturbance on (or interaction with) any conservative field surrounding it. The mentioned fields are produced by independent bodies and exist independently from the object, as mentioned previously.

5.1. 10. As mentioned in [5.1. 8], first and last points of ZT have equal value of S. In the real world, the value of S depends on the location of a particle and the bodies that produce conservative fields. As soon as the position of a body changes from one point of time to another, we have changes of S on any particular point of space according to a given period of time. If two different points of space have the same value of S at different points of time, it can be used as first and last points of Z-Trajectory. The following equation shows that mathematically.

$$S_1(x_1, y_1, z_1, t_1) = S_2(x_2, y_2, z_2, t_2) \qquad \text{(a)}$$

Where:
- S_1 and S_2 are values of S on two independent points of space

38

- x_1, y_1, z_1 is coordinates of first point of space
- x_2, y_2, z_2 is coordinates of second point of space
- t_1 is first point of time
- t_2 is second point of time

5.1. 11. Equation [5.1. 10. A] shows basic laws for first and last point of ZT. To calculate all possible number of points where the last point of ZT can exist, we need to find all the number of points where S equals the S of first point of ZT. Therefore, equation [5.1. 10. A] turns into the following one.

$$S_1(x_1, y_1, z_1, t_1) = S_2(x_2, y_2, z_2, t_2) = S_n(x_n, y_n, z_n, t_n) \qquad (a)$$

S_n is any point of space with coordinates x_n, y_n, z_n and points of time t_n where S_n equals to S_1.

Equation (a) shows the general law for ZT as follows.

5.1. 12. In case of the possible presence of a number of points in space with equal value of S, any two of them can be used as first and last point of Z-Trajectory.

5.1. 13. The previous statement can be shown graphically.

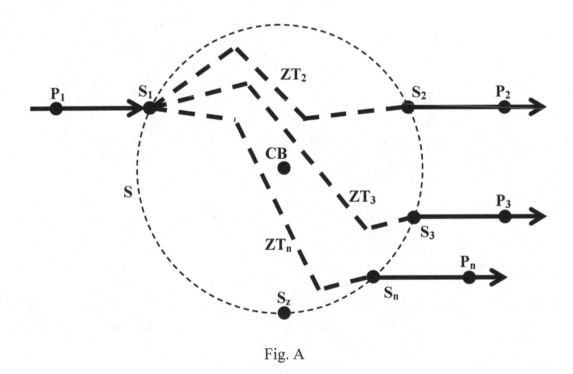

Fig. A

In figure [A], P_1 - S_1 is part of the RWT of a particle. S is number of points with equal S_g. Points S_2, S_3, and S_n are a few points from the countless number of such points on circle S. CB is the central body for circle S. ZT_2, ZT_3, and ZT_n are possible Z-trajectories for a particle. S_2 - P_2, S_3 - P_3, and S_n - P_n are possible RWT for a particle after its travel through ZT.

In the case of the presence of a gravity field surrounding the central body on figure [A], the whole case is described by equation [5.1. 6.A].

Chapter 6. Matter of Observation

6.1. 1. The previous chapter discussed very important aspects of Z-Trajectory. A key aspect is mentioned at [5.1. 9]. What do those aspects mean for an independent observer?

6.1. 2. in the case of a gravity field, the absence of interaction between an object on ZT and the gravity field surrounding it that is produced by an independent body leads to the inability of an independent observer to notice any movement of an object by means of the gravity field. That happens because an object produces no changes in the gravity field of the independent body. As a result, any motion of an object, as long as it exists on ZT, is hidden from gravity observation and any other type of observation where the gravity field is involved.

6.1. 3. Here and later I refer to the gravity field of an independent body as G-IB. The electric field of an independent body I refer to as E-IB. For other types of conservative field of independent body, I use C-IB.

6.1. 4. In the case of electric fields, absence of interaction between an object on ZT and E-IB leads to the inability of an independent observer to notice any movement of an object by means of an electric field. That happens because an object produces no changes on E-IB. As a result, any motion of an object, as long as it exists on ZT, is hidden from observation by electric the field and any other type of observation where electric field is involved.

6.1. 5. Paragraph [6.1. 4] leads us to a very important conclusion. If an object is unobservable by any type of electric field, it can't be observable by any type of electric field derivation, such as electromagnet waves, because any type of such wave contains an electric component. "A manipulation of the four equations for the electric and magnetic fields led Maxwell to wave equations for the fields, the solutions of which are traveling harmonic waves. Though the mathematical treatment is detailed, the underlying origin of the waves can be understood qualitatively: changing magnetic fields produce electric fields, and changing electric fields produce magnetic fields. This implies the possibility of an electromagnetic field in which a changing electric field continually gives rise to a changing magnetic field, and vice versa."[1]

6.1. 6. "*Magnetic field* is the region in the neighbourhood of a magnet, electric current or *changing electric field*, in which magnetic forces are observable."[2] Therefore, a magnetic field cannot exist without a *changing electric field*. Therefore, if nothing changes at the electric component of an electromagnet wave, nothing changes in the magnet component of that wave. Hence, the magnet component of electromagnet wave

---

[1] "Light." Encyclopaedia Britannica. Encyclopaedia Britannica 2008 Deluxe Edition. Chicago: Encyclopaedia Britannica, 2008.
[2] "Magnetic field." Encyclopaedia Britannica. Encyclopaedia Britannica 2008 Deluxe Edition. Chicago: Encyclopaedia Britannica, 2008.

is unable to have interaction with an object on Z-Trajectory. That leads us to following statement.

6.1. 7. As long as an object keeps Z-Trajectory, it cannot be observable by any type of electromagnet wave. In general consideration, we have the following conclusion.

6.1. 8. *As long as an object keeps Z-Trajectory, it cannot be observable by any disturbance or any interaction with any type of conservative fields, as well as by any derivation from those fields, including waves.*

6.1. 9. As a result we have no possible physical experiment to determine the presence or "location" of an object between first and last points of Z-Trajectory.

Chapter 7. Matter of Second Body

7.1 Equal Bodies

7.1. 1. Equation [5.1. 11.A] was used to determine the number of points around a central body with equal value of S between whom ZT become possible. It looks good for the ideal scenario when only the body is involved in the event. In the real world there are a lot of bodies producing gravity and other types of conservative fields. Is it possible to use that equation to describe the necessary state for general consideration? To answer that question, we need to analyze a case of two bodies and the motion of a particle between them.

7.1. 2. In the case of two bodies, we have an example of pair bodies with equal masses producing equal gravity fields.[1] That state is shown in the following figure.

Figure [A] shows two bodies CB_a and CB_b positioned on points A and D (letters above), and particle P moved from first position (P_a) to last position (P_b). Those positions mentioned are points B and C respectively. Points (A,B,C,D) lie on a straight line. AB is the radius for a circle drawn around point A, and DC is the radius for a circle drawn around point D. The length of line segment AB is equal to the length of line segment CD. As a result booth circles represent a number of points around central bodies with equal SGF.

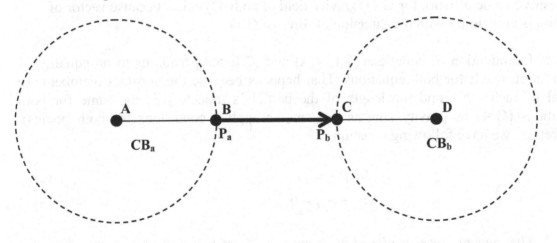

Fig. A

7.1. 3. As mentioned before at [2.3.5. A], the force produced on a unit particle by a body with mass m and distance between them r can be calculated by the following equation.

---

[1] This is a special case.

$$F_g = G \cdot \frac{m}{r^2}$$ (a)

7.1. 4. To calculate the work produced by the gravity field of body A to move a particle with unit mass from point B to point C, we can use equation [3.3. 1.A]. Because force of gravity field changes according to the distance between the particle and the body, we have the following equation for calculating the exact value of work.

$$A_{ga} = -\int_b^c G \cdot \frac{m_a}{r^2} \cdot dr$$ (a)

A negative value of work for G-A (gravity field of body A) exists because vector of motion is opposite to the direction of force of G-A.

7.1. 5. To calculate work produced by the gravity field of body D in order to move a particle with unit mass from point B to point C, we can use equation [3.3. 1.A]. Because force of gravity field changes according to the distance between the particle and the body, we have the following equation for calculation exact value of work.

$$A_{gd} = \int_b^c G \cdot \frac{m_d}{r^2} \cdot dr$$ (a)

A positive value of work for G-D (gravity field of body D) exists because vector of motion is coinciding with the direction of force of G-D.

7.1. 6. Examination of equations [7.1. 4.A] and [7.1. 5.A] leads us to an equality of calculation result for both equations. That happens because the constant members are equal to each other and the length of the particle's trajectory is the same for both equations (G=G as gravity constant; $m_a=m_d$ as applied conditions for both bodies). Therefore, we have following equations.

$$A_{ga} = -A_{gd}$$ (a)

$$A_{ga} + A_{gd} = 0$$ (b)

7.1. 7. *The sum of work produced by gravity fields of two bodies – in the case of a particle moving along a straight line and connecting the centers of those bodies from the point positioned on a given distance from the center of the first body to the point positioned on the same distance from the center of the second body – is equal to zero.*

7.1. 8. In general cases there is a more complex structure of particle motion in gravity fields of two bodies with equal masses. The figure shows such a case.

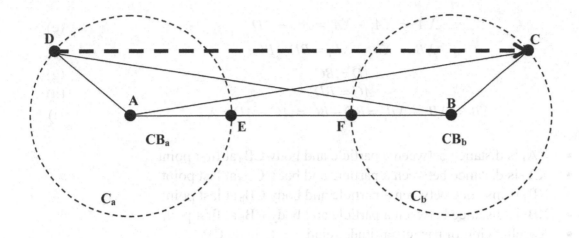

Fig. A

In figure [A] two bodies CB_a and CB_b are positioned on points A and B like figure [7.1. 2.A]. AB is a straight line that connects the centers of the bodies. AE=AD, so a number of points on circle C_a around A means a number of points with equal SGF. BF=BC, and as a result a number of points on circle C_b around B means a number of points with equal SGF. AE=FB like figure [7.1. 2.A], and as a result AD=AE=BF=BC.

Now draw a circle with its center on point B and some radius R that has the following rules: [R > AB] and [R < (AB+AD)]. In that case the circle built on R (now drawn on figure) crosses circle C_a on some point mentioned on the figure as point D. In that case the radius R equals BD.

With the same manipulation with circle C_b, same value of radius R, and the center of the circle on point A, it gives us point C on circle C_b. Therefore, R=AC=BD.

7.1. 9. As mentioned before at [3.5. 2], potential energy of a particle moving in a gravity field of a body depends only on the first and last points of a particle motion regardless of trajectory. Therefore, the only significant measure for calculation of changing a particle's potential energy is the distance between the first and last points of a particle moving and the central body. In other words, changing of potential energy depends only on the difference of altitude (distance) between a body and a particle, regardless of particle trajectory.

7.1. 10. Regarding figure [7.1. 8.A], it's quite possible to evaluate the changing of distance between body CB_a and a particle and body CB_b and a particle moving between points D and C.

$$XA_1 = AD \qquad\qquad\text{(a)}$$
$$XA_2 = AC \qquad\qquad\text{(b)}$$
$$XB_1 = BC \qquad\qquad\text{(c)}$$
$$XB_2 = BD \qquad\qquad\text{(d)}$$

45

$$XA = XA_2 - XA_1 = AC - AD \tag{e}$$

$$XB = XB_2 - XB_1 = BD - BC \tag{f}$$

$$AD = BC \tag{g}$$

$$AC = BD \tag{h}$$

$$XB = XB_2 - XB_1 = BD - BC = AC - AD = XA \tag{i}$$

Where:

- XA_1 is distance between a particle and body CB_a at first point
- XA_2 is distance between a particle and body CB_a at last point
- XB_1 is distance between a particle and body CB_b at last point
- XB_2 is distance between a particle and body CB_b at first point
- XA changing of particle altitude relatively to body CB_a
- XB changing of particle altitude relatively to body CB_b
- Equations G and G are the same that was mentioned in [7.1. 8]

7.1. 11. Equation I is a result of equations A through H. That means in case of what is shown on figure [7.1. 8. A], the changing of altitude of a moving particle relative to the first body is equal to the changing of altitude of a moving particle relative to the second body. Therefore, the changing of potential energy of a moving particle relative to the first body equals the changing of potential energy relative to second body.

7.1. 12. Moreover, the gravity field of the first body produces negative work (and changes potential energy negatively) for the particle because its altitude increases during motion (AC > AD) [s. 7.1. 8. A]. Unlike the gravity field of the first body, the second one produces positive work (and changes potential energy positively) because the particle decreases its altitude during motion (BC < BD) [s. 7.1. 8. A].

7.1. 13. Therefore, the first and second body changes potential energy of a moving particle between points D and C at the same value but with different signs. As a result, such movement of a particle changes no energy of the whole system by the gravity fields, including both bodies and the particle. The following equations shows that mathematically.

$$\sum P(D) = \sum P(C) \tag{a}$$

Equation (a) leads us to following conclusion.

7.1. 14. As shown at [3.2. 4], SGF depends only on the distance between any particular point and a body producing a gravity field (in the case of constant mass of the body). According to figure [7.1. 8.A], the SGF of points D [$S_g(D)$] and C [$S_g(C)$] can be calculated the following way:

$$S_g(D) = G \frac{m_a}{(ad)^2} + G \frac{m_b}{(bd)^2} \tag{a}$$

$$S_g(C) = G\frac{m_a}{(ac)^2} + G\frac{m_b}{(bc)^2} \qquad\qquad (b)$$

Because $m_a = m_b$ [s.7.1. 8] and [7.1. 10. g–h], we have following equation:

$$S_g(D) = G\frac{m}{(ad)^2} + G\frac{m}{(bd)^2} = G\frac{m}{(ac)^2} + G\frac{m}{(bc)^2} = S_g(C) \qquad (c)$$

That leads us to following conclusion.

7.1. 15. In the case of a moving particle in gravity fields of two bodies with equal masses, the sum of the potential energy of that particle keeps constant when the motion produced between the first and last points are equal SGF (S_g).

7.1. 16. Equation [7.1. 14.C] shows the calculation of gravity force magnitude on two particular points of space. Because any force has direction and magnitude, the sum of two vectors is a vector. That vector depends on the directions of the two vectors' sum. In the case of [7.1. 8], we have two vectors for point D and C. Magnitudes of those vectors are calculated by [7.1. 14.a–b] as members of the right part of each equation. Equation [7.1. 14.c] means an equality between vectors producing force along directions AD and BC and other pairs of equal vectors producing force along directions BD and AC.

7.1. 17. Because we need to evaluate the angle between each pair of vectors to evaluate their sum, we need to evaluate the angles of ADB and BCA of triangles ABC and ABD (ΔABC and ΔABD). Those triangles have one side, AB, that belongs to both triangles, and it has two pairs of independent sides, AD & BC and BD & AC. As mentioned in [7.1. 10.g-h], AD=BC and AC=BD. Therefore, we have the following sides of two triangles with equal length: AB=BA, AD=BC, AC=BD. Therefore, those triangles are congruent because of the equality of their three sides.

7.1. 18. In two congruent triangles, any element of the first one is equal to the appropriate element of the second one. Therefore, angle ABD=BAC, BAD=ABC, and ADB=BCA. Most important to us is the equality of two angles ADB and BCA because their sides coincide with the directions of gravity forces of bodies CB$_a$ and CB$_b$ on points D and C.

7.1. 19. Therefore, we have two pairs of equal vectors for each point D and C, and an equal angle between them at each point. In that case the vectors of sum for both points become equal to each other. *Therefore, the summary vectors of SGF on both points are equal to each other.*

7.1. 20. Because of this, equation [5.1. 5.A] is applicable for the case of two bodies as well as for the case of a single body.

7.1. 21. In case of other types of conservative fields, we have the same situation with only a difference on magnitude of forces. For example, in the case of an electric field, figure [7.1. 8.A] is correct: CB_a and CB_b are two bodies with the same electric charge (++ or --), and the particle moving from point D to point C carries little electric charge of any sign.

7.1. 22. In the three-dimensional world, we can rotate triangle ABC (fig. [7.1. 8.A]) around the AB axis. With that rotation, vertex C draws an exact circle (C-circle). Because the shape of triangle ABC has no changes, equation [7.1. 13.A] will be correct for any point of C-circle. The following figure shows that case. C-circle is drawn there as its projection (C_1-C_2) on the figure plane.

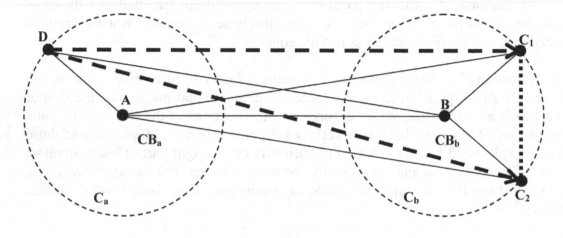

Fig. A

7.1. 23. Hence, at any motion of any particle by any trajectory that begins on point D and ends on any point of C-circle, the gravity fields change no energy of the moving particle as well as the whole energy on the system. That happens despite the different distance between point D and different points of C-circle, unlike the case of a single body where distance between first and last points of trajectory with equal SGF exists with the same remoteness from the central body.

7.1. 24. In the reverse case, if we rotate triangle ABD around the AB axis, vertex D draws an exact circle (D-circle). Because the shape of triangle ABD has no changes, equation [7.1. 13.A] is correct for any point of D-circle. The following figure shows that case. D-circle is drawn there as its projection (D_1-D_2) on the figure plane.

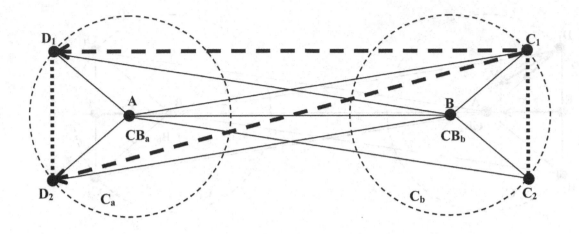

Fig. A

7.1. 25. Hence, at any motion of any particle by any trajectory that begins on point C and ends on any point of D-circle, the gravity fields change no energy of the moving particle as well as the whole energy on system. That happens despite the different distance between point C and different points of D-circle, unlike the case of a single body where distance between first and last points of trajectory with equal SGF exists with the same remoteness from the central body.

7.1. 26. The same consideration can be applied for any point of C-circle and D-circle. Hence, any point on them has equal value of SGF. Here and further I refer to such numbers of points as S-Outline.

7.1. 27. S-Outline means the number of points with equal value of S [s.5.1. 5.A]. Conservative fields change the energy of a particle by any motion of that particle between any numbers of such points by any trajectory so that the energy of a particle located at first point is always equal to its energy at last point. As a result, the relocation of a particle changes no energy of the whole system, including both bodies and the particle itself.

7.1. 28. In the case of single body, such points form an exact circle around the central body with any given radius R. In the case of two or more bodies, we have a more complex number of points with different distances between them and the central body, as shown by [7.1. 24]. As a result the plane radius becomes S-Outline.

7.1. 29. Looking closer at figure [7.1. 24.A], we have the following case for vectors that are shown on the following figure [A].

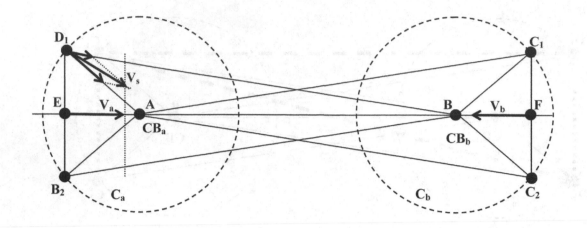

Fig. A

In the figure we have vectors V_s V_{bd}, and V_{ba} beginning on point D_1, and vector V_a as a projection of V_s on an extension of a straight line connecting the centers of bodies CB_a and CB_b. Vector V_{bd} means the force acting along line D_1B and represents the force produced from body CB_b on a particle with unit mass positioned on point D_1. Vector V_{ba} means the force acting along line D_1A and represents the force produced from body CB_a on a particle with unit mass positioned on point D_1. Vector V_s is the sum of vectors V_{bd} and V_{ba}. It represents the force produced by both bodies (CB_a and CB_b) on a particle with unit mass positioned on point B_1. The equation shows that mathematically.

$$\overline{V}_s = \overline{V}_{bd} + \overline{V}_{ba}$$ (a)

7.1. 30. Because of congruent triangles ABD_1 and ABC_1 [s.7.1. 17], and equal masses of central bodies CB_a and CB_b, we have the same number of vectors with the same orientation and magnitude on point C_1 (not shown on figure [7.1. 29.A]). Projection of their resultant vector on an extension of straight line AB gives us vector V_b.

7.1. 31. As a result vectors V_a and V_b lie on one straight line and have the same magnitude and the opposite direction. Therefore, the sum of those vectors equals zero. Here and further I refer to such vectors as Strength Projecting vectors (SP-vectors). Hence, the rule for S-Outline [s.7.1. 27] can be formulated the following way.

7.1. 32. In case of two bodies S-Outline means the number of points with equal value of field strength (S) [s.5.1. 5.A]. Those points are positioned in space so as the sum of projections of the resulting vectors on the one straight line connecting centers of two bodies equals zero. Projection on an addition of that line is acceptable too. Resulting vectors are formed by the sum of elemental vectors of field strength produced by two bodies on those exact points.

7.1. 33. The previous paragraph gives us a stronger formulation than the one mentioned above at [4.2. 7]. Moreover, it shows the same result used to find S-Outline by means of vector quantities as well as their magnitudes. As a result we have the following notion for first and last points of Z-Trajectory.

7.1. 34. *First and last points of Z-Trajectory are able to exist only on a S-Outline.*

7.1. 35. As we can see from figure [7.1. 29.A], S-Outline has two shapes, each of which are positioned closer to one body and farther from the other one. Here and later I refer to those parts of S-Outline as SA-Outline (the number of outline points located closer to body A) and SB-outline (the number of outline points located closer to body B). Hence, the full S-Outline can be described as the summation of the number of points of SA-Outline and SB-Outline.

7.2 Host Body

7.2. 1. Topic [7.1] explains an ideal case for Z-Trajectory for two bodies. In the real world there are a lot of bodies producing conservative fields of the same kind (G-IB and E-IB [r.6.1. 3]). Is it possible to apply the conclusions of [7.1] to the real world? This topic explains a possible way for such application.

7.2. 2. Paragraph [7.1. 29] shows vectors of S that lead to a general rule for S-Outline. Looking closer at figure [7.1. 29.A], we can understand the following details. If there is any additional force of any kind that adds vectors of that force to points D_1 and C_1, then those forces change magnitude and direction of vector V_s (resulting vector of S).

7.2. 3. Each additional body produces its own additional vector on both points (D_1 and C_1) [s.7.1. 29.A]. Because any number of vectors can be replaced by the resulting number of forces (one vector), we can use one additional vector applied to each point. Generally the magnitude and direction of those vectors are unequal.

According to the modern view on summarizing vectors, we have the following information. "Newton's law of gravitation and Coulomb's electrostatic law both give the force between two particles as inversely proportional to the square of their separation and directed along the line joining them. The force acting on one particle is a vector. It can be represented by a line with arrowhead; the length of the line is made proportional to the strength of the force, and the direction of the arrow shows the direction of the force. If a number of particles are acting simultaneously on the one considered, the resultant force is found by vector addition; the vectors representing each separate force are joined head to tail, and the resultant is given by the line joining the first tail to the last head."[1]

7.2. 4. Among the many cases described at [7.2. 3], there is one case where the sum of SP-vectors of those forces [s.7.1. 31] equals zero. As a result their presence or absence changes nothing on the projection of vector of force V_s on straight line AB [s.7.1. 29.A]. Therefore, in such a case, the presence of any number of extra bodies changes nothing in notion of [7.1. 34].

---

[1] "Physical science, principles of." Encyclopaedia Britannica. Encyclopaedia Britannica 2008 Deluxe Edition. Chicago: Encyclopaedia Britannica, 2008.

7.2. 5. What is the case where resulting vectors of additional forces produce forces with equal magnitude and at the same time produce equal projection of resultant forces on a straight line connecting two bodies? Projection on an addition of that line is acceptable too. The easiest way to imagine such a state is this. There is one body with any exact mass that produces equal S_c on each pair of S-Outline points. If that body is located an equal distance from first and second body (CB_a and CB_b [s.7.1. 29.A]), it becomes possible. Here and further I refer to such a body as Host Body (HB).

7.2. 6. HB can be imagined as a body that produces equivalent C-field as described in [7.2. 5] or a real body producing the same field (fields). Both cases lead to the same result: the possible case of the presence of S-Outline for two bodies in the real world (there are a lot of different bodies and fields that exist simultaneously).

7.2. 7. The presence of HB is possible despite the value of its mass because each body producing C-field has the same value of S_c on any point equidistant from the body.

7.2. 8. Which position is possible for HB? The number of points that are equidistant from two exact points in a plane lies between those points perpendicular to straight line that connects those two points in the very middle of that line.[1] Here and further I refer to that plane as Host Body Equal Strength plane (HBES-plane) Therefore, as long as HB positions itself on any exact point of HBES-plane, all sophisticated reasoning mentioned in [7.1–7.2] are correct.

7.3 Unequal Bodies

7.3. 1. In the case of two bodies producing unequal conservative fields, figure [s.7.1. 29.A] can be redrawn as follows.

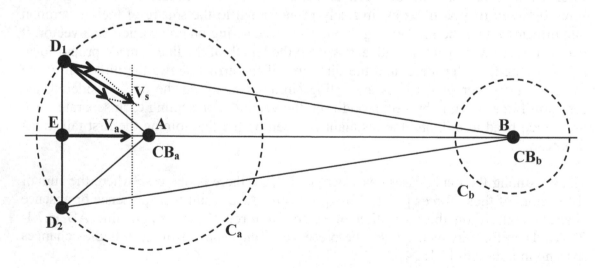

Fig. A

---
[1] From stereometry.

7.3. 2. Figure [7.3. 1.A] shows two bodies CB_a and CB_b producing unequal conservative fields. In the case of gravity fields, those bodies have different masses. Circles C_a and C_b represent the number of points with equal S_c (in the case of gravity field $S_c = S_g$). As you can see, the radius of C_b is less than the radius of C_a. That represents a weaker field produced by body CB_b. In the case of gravity fields, the mass of CB_b is less than the mass of CB_a.

7.3. 3. Because of the difference between two bodies, we can't find SB-Outline as it was done previously [7.1]. The key difference between those two cases is that the unequal fields have a dependence of distance between the bodies. It happens because static fields produce unequal work with equal movement of a particle in the case of unequal fields. Therefore, a symmetric S-Outline [Fig. 7.1. 24. A] is impossible.

7.3. 4. Paragraph [7.1. 30] is not applicable in that case because first and second bodies produce different forces of the static field at the same distance from them. Hence, there is only one possible way for the existence of S-Outline as a number of points with equal distance from the second body. All those points have same value of S_c and lie on the circle mentioned in figure [7.1. 24. A] as its projection on figure plane (D_1-D_2). That circle is SA-Outline, as was already mentioned.

7.3. 5. **Conclusion**. In the case of two bodies producing unequal conservative fields, S-Outline becomes SA-Outline.

7.3. 6. Conclusion [7.3. 5] is applicable to both bodies. If we swap bodies, replacing CB_a by CB_b and vice versa [s. 7.3. 1. A], we have the same figure with the only difference being that the first body produces a lesser field than the second one. Hence, [7.3. 4] is applicable in that case too.

7.3. 7. In the case of two bodies producing unequal conservative fields, Z-Trajectory is only possible between any points of SA-Outline [b. 7.3. 1–7.3. 6].

7.3. 8. In the real world we can replace all vectors of gravity force interacting with a particular object by the resultant of its forces. As a result, all gravity forces produced by any part of the universe can be replaced by their resultant forces, and the whole interaction between the object (located not far from the body) and the universe can be reduced to figure [7.3. 1.A]. In that case the whole mass of the universe is replaced by one body with a mass equal to some particular mass that produces the same S_g on the area where the object (and the body interacting with it) is located (the position of that body is shown as location CB_b on figure [7.3. 1.A]).

7.3. 9. The previous paragraph explains the possibility of SA-Outline's presence around any real body in universe. Therefore, as long as equation [5.1. 11.A] is correct, the presence of SA-Outline in the real world is possible as a specific number of points in space.

Chapter 8. Matter of Energy

8.1 Law of Conservation

8.1. 1. In modern physics there are a number of strong law of conservation. Generally they represent the result of theories passed down by physics from the ancient days to present time. Those laws are basic principles for all modern natural sciences. Among them there is the main law of the conservation of energy. The following paragraph shows that law.

8.1. 2. "Conservation of energy implies that energy can be neither created nor destroyed, although it can be changed from one form (mechanical, kinetic, chemical, etc.) into another. In an isolated system the sum of all forms of energy therefore remains constant. For example, a falling body has a constant amount of energy, but the form of the energy changes from potential to kinetic. According to the theory of relativity, energy and mass are equivalent. Thus, the rest mass of a body may be considered a form of potential energy, part of which can be converted into other forms of energy."[1]

8.1. 3. That law plays a very important role in any new theories and experiments because it gives a strong key to check the consistency of any theory or experiment. If an experimenter observes something that violates the conservation of energy law, that leads to an understanding of missing something during the experiment. An excellent example of using that method is the discovery of the neutrino.

8.1. 4. "In 1928 Pauli became professor of theoretical physics at the Federal Institute of Technology, Zurich. Under his direction the institution became a great centre of research in theoretical physics during the years preceding World War II. In the late 1920s it was observed that when a beta particle (electron) is emitted from an atomic nucleus, there is generally some energy and momentum missing, a grave violation of the laws of conservation. Rather than allow these laws to be discarded, Pauli proposed in 1931 that the missing energy and momentum is carried away from the nucleus by some particle (later named the neutrino by Enrico Fermi) that is uncharged and has little or no mass and had gone unnoticed because it interacts with matter so seldom that it is nearly impossible to detect. This particle was not observed until 1956."[2]

8.1. 5. "The basic properties of the electron-neutrino – no electric charge and little mass – were predicted in 1930 by the Austrian physicist Wolfgang Pauli to explain the apparent loss of energy in the process of radioactive beta decay. The Italian-born physicist Enrico Fermi further elaborated (1934) the theory of beta decay and gave the 'ghost' particle its name. An electron-neutrino is emitted along with a positron in

---

[1] "Conservation law." Encyclopaedia Britannica. Encyclopaedia Britannica 2008 Deluxe Edition. Chicago: Encyclopaedia Britannica, 2008.
[2] "Pauli, Wolfgang." Encyclopaedia Britannica. Encyclopaedia Britannica 2008 Deluxe Edition. Chicago: Encyclopaedia Britannica, 2008.

positive beta decay, while an electron-antineutrino is emitted with an electron in negative beta decay."[1]

8.1. 6. As we can see from [8.1. 4], Mr. Pauli believes the law of conservation despite the particular experimental result. Because he was professor of theoretical physics, he believes as well that if there is some disparity between theoretical law and the results of an experiment, real problems exist in the experimental area, not in the theoretical one. Such a point of view led him to the understanding of absence of wholeness of an experiment that was shown by missing something that carried away[2] some energy.

8.1. 7. Going further in that direction we need to make the next step and accept this. As well of wholeness of experiment, we need to use completeness of experiment to use the conservation law as a certain law of the universe. The difference between wholeness and completeness of experiment lies in the unlikeness between dimension and time. In other words, wholeness of experiment relates to any aspect of experiment in space, and completeness of experiment relates to any aspect of experiment in time.

8.1. 8. Here is an example for [8.1. 7]. In the case of radioactive beta decay, an experimenter who observes that process has very little chance to detect neutrinos in the same laboratory where beta decay occurs. But it's quite possible to detect such a particle from beta decay happening at some distant place and moved to a laboratory, after some time that is needed for the relocation of a particle from its point of creation to the point of detection. Hence, there is a time interval between the event (beta decay) and the observation of a completed event, which includes creation and observation (destruction) of a particle.

8.1. 9. Hence, we have the following notion for the law of conservation.

8.1. 10. Conservation of energy implies that energy can be neither created nor destroyed, although it can be changed from one form (mechanical, kinetic, chemical, etc.) into another. In an isolated system, the sum of all forms of energy therefore remains constant on any whole and competed event.[3]

8.1. 11. With the law of conservation according to [8.1. 10], we can use it as a measure of energy and as a measure of wholeness and completeness of any event. That way was how it was successfully used firstly by Mr. Pauli in 1931. But a stronger law (as [8.1. 10]) was not introduced at that time. Generally that happened because any experimenter believes that he provides wholeness and completeness of an experiment on his laboratory a priori'. As a result each case whose aspects are violated leads to the idea of "incorrectness" of the law of conservation.

---

[1] "Neutrino." Encyclopaedia Britannica. Encyclopaedia Britannica 2008 Deluxe Edition. Chicago: Encyclopaedia Britannica, 2008.
[2] By means of some process that cannot be noticed by an experimenter.
[3] In that notion, "event" and "experiment" look equal because "experiment" is one of the "events" happening continuously in the world.

8.1. 12. Law of conservation in the notion of [8.1. 10] can be easily used to check the experiment itself and its surrounding environment. Each time an experimenter observes a "violation" of the fundamental law of conservation, the person misses something in space or time according to the aspects of any given experiment.

8.1. 13. Usually that happens in some experiments related to areas that are not fully researched and described. Those areas look like the edge of science, where our knowledge and imagination cannot be quite correct. As a result it's very possible for an experimenter to "see" something away from law of conservation. Here and further I refer to any of such experiment as an Edge-experiment.

8.1. 14. Now we can see that there is no way to violate the law of conservation.

8.2 Conservation and Observation

8.2. 1. As shown before at [8.1], any aspect of an experiment (event) must be observable by the experimenter to fit the law of conservation. Any deviation from that rule causes a so-called violation of a basic principle.

8.2. 2. Historically the law of conservation became understandable only when experimenters were able to observe most aspects of their experiments. Each time when whey found all features of experiment and understood the balance of energy, they shared the law of conservation on that experiment. That way led physicians to the understanding of applicability of a basic principle to all possible events in our world. Today that principle is well-known as law of conservation.

8.2. 3. At some time physicians noticed a "violation" of the law of conservation in this or that experiment. In modern physics, that matter is usually solved by closer observation of the experiment with the aim to find a "missing" aspect of the experiment (mass, energy, etc.).

8.2. 4. But that is only possible in cases of observable aspects. In other words, an aspect must be observable in some way to be recognized as part of the experiment in order to build a clear view on the wholeness of the experiment [s. 8.1. 4].

8.2. 5. Therefore, any case when we have something unobservable on any experiment (event) leads us to assume a violation of observation, not a violation of conservation.

8.3 Energy and Z-Trajectory

8.3. 1. Any experiments (events) where experimenters miss not only some matter or energy but a whole object drives modern physics to nonplus. But if you remember chapter 6 of this work, it must be clear to the scientist to see a difference between undetectable objects and nonexistent objects.

8.3. 2. Hence, the presence of an object in Z-Trajectory, without any way to be observed [s. 6], is not a violation of the law of conservation because the object still exists regardless of the possibility of its observation. Moreover, the law of conservation gives way to an object to keep Z-Trajectory. Let's see how it is possible.

8.3. 3. The law of conservation can be expressed mathematically by the following equations.

$$\sum E_b = \sum E_a \qquad \text{(a)}$$
$$\Delta E = \sum E_b - \sum E_a = 0 \qquad \text{(b)}$$

Where:
- ΣE_b is sum of energy on isolated system before process (experiment, event, etc.)
- ΣE_a is sum of energy on isolated system after process (experiment, event, etc.)
- ΔE is difference between sum of energy before and after process (experiment, event, etc.)

Equation (a) represents the law of conservation itself. It shows equality between the amount of energy on an isolated system before and after an experiment (event).

Equation (b) represents the difference between the amount of energy of an isolated system before and after the experiment (event). It is calculated as the difference of energy in an isolated system before and after the experiment (event). It always equals zero because of the equality of energy before and after the experiment. That leads us to a very important conclusion.

8.3. 4. *The law of conservation permits the presence of only processes that bring no energy in or out of an isolated system.*

8.3. 5. As shown at [4.2. 6], the general law of Z-Trajectory means the same thing. Any particle that is displaced by Z-Trajectory has the same amount of energy at its first and last points. That is the key property of Z-Trajectory that permits it to exist according to [8.3. 4].

8.3. 6. Changing of energy of a particle moving by Z-Trajectory exactly matches equation [8.3. 3.b] because of [4.2. 3.a][1]. Therefore, the present condition of Z-Trajectory itself is not a contradictory condition for the law of conservation (and [8.3. 4] as well).

---

[1] Chapters 5 and 7 give detailed explanation for [4.2.3].

8.4 Energy and Observation

8.4. 1. It's time to go back a bit and refresh our memories about something discussed in chapter 6. That chapter described the absence of possibility for an object to be observed in any way as long as it keeps Z-Trajectory.

8.4. 2. You may have asked yourself how it is possible to have something without its interaction with the surrounding world. In the case of observation of any physical process, we have more or less information about the process itself, its features, and the stage of the surrounding mater. Experimenters have such information due to the interaction between an experimenter and the observing object.

8.4. 3. For example, an experimenter watches any object the person interacts with it by means of light waves. Light waves must be emitted by any source and spread across some distance to interact with the object and travel to the person. That's the only possible way for optical observation (e.g., by human vision).

8.4. 4. Generally any observation means an interchange of some energy between the observing object and the observer.

8.4. 5. Moreover, any interaction between an object and an observer (not only observation) involves some energy interchange between them. In the case when there is no energy interchanges between the object and the observer, no interaction is possible.

8.4. 6. The only way that gives the object(s) opportunity to reach such a condition is the law of conservation.

8.4. 7. That law mentioned the notion of isolated system [s. 8.1. 2]. Isolated systems in modern physics means any number of physical units[1] that have energy interaction only between themselves. As a result there is no energy going in or out of such a system.

8.4. 8. To separate an isolated system as an imaginable system and a physical system in a state where isolation is the main aspect of that system, I use the notion "closed state" for any type of such systems (PS_c).

8.4. 9. For a physical system in its usual state, when its units have interaction with the surrounding world and any other physical unit, I refer to it as physical system in open state (PS_o).

8.4. 10. Hence, the difference between physical systems in closed and open states lie only in the area of interaction between units of the system and the surrounding world.

8.4. 11. A system in a closed state permits energy interchange only between its units.

---

[1] Particles, fields, etc.

8.4. 12. Hence, there is no way for energy interaction between units of a system in a closed state and any physical unit located outside of the system. Therefore, there is no way to have any information about the system in a closed state if one is outside of the system.

8.4. 13. Furthermore, any system in a closed state is unable to have any interaction with any observer located outside of the system. In other words there is no possible physical process (experiment, event, etc.) that can show the presence or absence of any system in such a state anywhere.

8.5 Energy and Systems in Different States

8.5. 1. To keep the law of conservation before, in, and after the system using Z-Trajectory, it must retain following rules.

    a.   System must keep the law of conservation before Z-Trajectory
    b.   System must keep the law of conservation going into Z-Trajectory
    c.   System must keep the law of conservation inside Z-Trajectory
    d.   System must keep the law of conservation going out of Z-Trajectory
    e.   System must keep the law of conservation after Z-Trajectory

8.5. 2. The rules mentioned above [8.5. 1.a–e] can be shown by following figure.

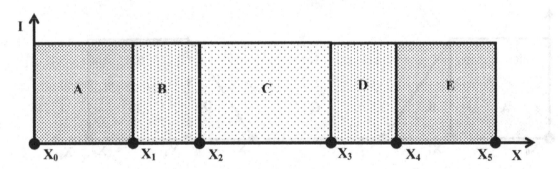

Fig. A

Figure [A] shows a number of states of a physical system (PS) that uses Z-Trajectory as part of its whole trajectory (RWT).

A, B, C, D, and E are states of PS. Axis X represents the motion of PS in space. Axis I represents the level of interaction of units of the system between themselves.

8.5. 3. As we can see, the level of interaction between inner units of the system in the open and closed states are equal to each other. A sequence of locations for the system mentioned on figure [8.5. 2.A] means following:

a. A is the location of the system before Z-Trajectory (number of points between X_0 and X_1)
b. B is the location of the system going to Z-Trajectory (number of points between X_1 and X_2)
c. C is the location of the system inside Z-Trajectory (number of points between X_2 and X_3)
d. D is the location of the system coming from Z-Trajectory (number of points between X_3 and X_4)
e. E is the location of the system after Z-Trajectory (number of points between X_4 and X_5)

8.5. 4. Because of the absence of changes between levels of interaction between inner units of the system on *any* location (before, in, out, of ZT), the energy interaction between them remains unchanged. That aspect has the following consequence on the law of conservation: it keeps it unaffected for any system in a closed state.

8.5. 5. As a result there is no physical process inside the system that can be changed by isolating the system.

8.5. 6. The only process that is affected by isolating is the process involved in the interaction between inner units of the system and the surrounding world (and vice versa). The following figure shows the interaction changing between units of the isolating system and the surrounding world.

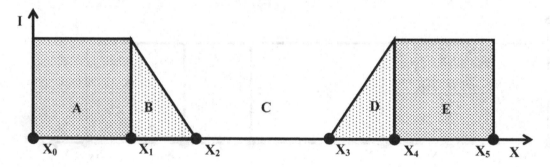

Fig. A

8.5. 7. Elements of figure [8.5. 6.A] have the following meaning.

a. A, B, C, D, and E are states of PS. Axis X represents the motion of PS in space. Axis I represents the level of interaction between units of the system between inner units and units of surrounding world (and vise versa).

b. A is the location of the system before Z-Trajectory (number of points between X_0 and X_1). At that location, the system has full interaction between inner units and units of the surrounding world (and vice versa). As a result the state of the system is no different from its usual condition.

60

c. B is the location of the system going into Z-Trajectory (number of points between X_1 and X_2). This is a very special location with the changing level of interaction between inner units of the system and units of surrounding world (and vise versa). The closer the system goes to X_2 point, the less level of interaction between units of the system and units of surrounding world.

d. X_2 is the point where level of interaction between inner units of the system and units of surrounding world (and vice versa) become zero.

e. C is the location of the system where the interaction between inner units and units of the surrounding world (and vice versa) equals to zero. *That is Z-Trajectory itself.* As mentioned above, there is no possibility to observe a system in such a state because of the absence of energy interaction with the system from the outside (and vice versa).[1]

f. X_3 is the point where the level of interaction between inner units of the system and units of surrounding world (and vise versa) starts to rise above zero.

g. D is the location of the system going from Z-Trajectory (number of points between X_3 and X_4). This is a very special location with the changing level of interaction between inner units of the system and units of surrounding world (and vice versa). The closer the system goes to X_4 point, the higher the level of interaction between units of the system and units of surrounding world.

h. E is the location of the system after Z-Trajectory (number of points between X_4 and X_5). At that location the system has full interaction between inner units and units of the surrounding world (and vice versa). As a result the state of the system is no different from its usual condition.

8.5. 8. Generally A and E are the locations of the system with its usual state. They have no difference between the usual state of space and any system positioned where there is the usual level of interaction with the surrounding world. That happens because any amount of energy that is sent to the system is accepted by the system completely according to the physical properties of its units. As a result any process of interaction between units of the system and surrounding world stays unchangeable.

8.5. 9. Locations B and D have only some level of interaction between units of the system and surrounding world. As a result only *part* of the energy that is sent to the system from the outside reaches units of the system (and vise versa). The less energy able to reach units of the system, the less level of interaction between the system and surrounding world.

---

[1] Looking from the system located outside of Z-Trajectory.

Chapter 9. Neighborhood of Zero

9.1 Zero Interaction

9.1. 1. The figures mentioned in the previous chapter [s. 8.5. 2.A; 8.5. 6.A] show two opposite cases, the first of which shows full interaction between units of same system located on zoon C, and the second shows zero level of interaction between units of different systems with one of them located on zoon C.

9.1. 2. This brings up a very important question: is it necessary to change level of interaction between interacting systems, or is it enough to change the interaction only between some areas in space to reach same result?

9.1. 3. Usually any type of interaction includes some energy interchange between units of different systems. In the case of energy interchange between units of the same system, we have inner interaction between units of same system. Hence, there is only one important subject of energy interchanged between units regardless of the systems to which they belong.

9.1. 4. The type of interaction between two physical units depends on their properties. As well as any property changes, interaction between those units itself changes according to the modifications of those properties. But interaction still exists because of the possibility of energy interaction between physical units.

9.1. 5. No interaction appears only when there is no possibility of energy interaction between physical units. In the case of the easiest mechanical interaction between two mechanical units, such a case can be reached by the modification of object properties. For example, we can separate two mechanical units[1] and break any mechanical interaction and mechanical energy intercourse between them.

9.1. 6. There is a different situation in the case of field units. There is no way to break fields' interaction between themselves as well as between mechanical objects. To reach this result, we need to separate them in space to break the interaction between fields and objects or between two or more fields.

9.1. 7. Hence, isolation of some space gives us the possibility to break interaction between field units of a system located in and out of isolated space. Therefore, in the case of reducing interaction between all units of any type of one system and a second system, we need to have a separating area. Such an area must be the most distant place that is reachable for interaction (for energy interaction by fields). Here and later I refer to such area as Event shield (E-Shield) because observation of any event is impossible outside of it (behind the shield).

---

[1] The separable property that involves the distance between two objects.

9.1. 8. **Conclusion.** There are two different areas separated by the E-Shield with the following characteristics.

   a. Each area is unable to have any interaction (energy intercourse, information interaction, etc.) with another one
   b. Each area looks like a closed system from the other one
   c. Each area follows the law of conservation (for any physical interaction inside it)

Therefore, any unit located in space is able to interact only with units on the same side of the E-Shield regardless of the number of units and the types of their interaction. In other words, a single E-Shield separates space on two isolated areas. Each area is able to contain any number of physical units, allowing interaction between them according to the law of conservation.

9.1. 9. Here and further I refer to two zones separated by an E-Shield as CE-Space (Clear Event Space) and HE-Space (Hidden Event Space).

9.1. 10. Obviously, it's a relative identification of spaces. For any IO there is only the surrounding CE-Space. It is impossible for the IO to see HE-Space and anything located there. But any IO located in a different space (HE-Space from the point of view of first IO) has CE-Space in the area where the first one has HE-Space, and vice versa.

9.2 Reduced Interaction

9.2. 1. The previous topic [9.1] discusses aspects of a fully established E-Shield. In such cases there is only interaction between objects located on CE-Space. HE-Space is unreachable for interaction as well as any physical unit located there.

9.2. 2. Hence, on the one hand we have at least two spaces separated by E-Shield without any interaction between them. On the other hand the separated spaces have no difference in interaction between the physical units belonged to them. That leads us to the following questions.

9.2. 3. Is it possible for the E-Shield to exist at one state?[1] Is it possible to have some (variable) level of separation and interaction? In such cases an interaction diagram can be drawn.

---

[1] Full separation.

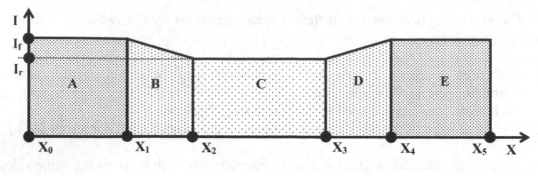

Fig. A.

9.2. 4. Elements of figure [9.2. 3.A] have the following meaning.

a. A, B, C, D, and E are states of PS. Axis X represents motion of PS in space. Axis I represents levels of interaction of the units of the system with the inner units and units of the surrounding world (and vice versa).

b. A is the first location of the system (number of points between X_0 and X_1). At that location the system has full interaction (I_f is level of full interaction) between inner units and units of surrounding world (and vice versa). As a result the state of the system is no different from its usual condition.

c. B is the location of the system going to some space with a reduced interaction (number of points between X_1 and X_2). This is a very special location with a changing level of interaction between inner units of the system and units of surrounding world (and vice versa). The closer system goes to X_2 point, the less level of interaction between units of the system and units of surrounding world.

d. X_2 is the point where the level of interaction between inner units of the system and units of surrounding world (and vice versa) reaches its lowest value (I_r).

e. C is the location of the system where interaction between inner units and units of surrounding world (and vice versa) is equal to the lowest value. Energy interchange and interaction between units of the system and surrounding world on that area equals I_r (reduced level of interaction).

f. X_3 is the point where the level of interaction between inner units of the system and units of surrounding world (and vice versa) starts to rise above the minimal level of interaction (I_r).

g. D is the fourth location of the system (number of points between X_3 and X_4). This is a very special location with changing levels of interaction between inner units of the system and units of surrounding world (and vice versa). The closer the system goes to X_4 point, the more interaction that exists between units of the system and units of surrounding world.

h. E is the fifth location of the system (number of points between X_4 and X_5). On that location the system has full interaction between inner units and units of the surrounding world (and vice versa). As a result the state of the system is no different from its usual condition.

9.2. 5. As we can see from [9.2. 3.A], the more difference between the level of full interaction (I_f) and the reduced one (I_r), the closer the space (between X_2 and X_3) to a state of a closed system. As soon as I_r becomes zero, X_2-X_3 space becomes HE-Space looking from any other area mentioned on the figure.

9.3 Interaction and Shield

9.3. 1. Topics [9.1; 9.2] discussed the interaction looking from our side of space. It's time to discuss interaction from the E-Shield side.

9.3. 2. Interaction between physical units located in CE-Space and HE-space is shown on the following figure.

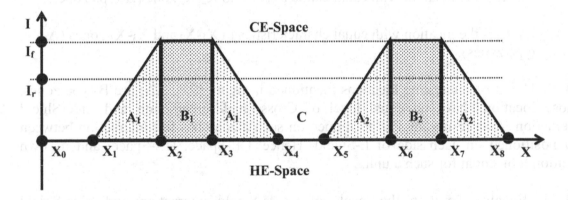

Fig. A

9.3. 3. Elements of figure [9.3. 2.A] have the following meaning.

a. A_1, B_1, A_2, B_2 are states of interaction between CE-Space and HE-Space (and vice versa). That condition represents interaction between physical units (PU) located at different points of space (X axis). Here and later I refer to such interaction as Cross-Shield interaction. As mentioned above, the level of that interaction is usually equal to zero. Axis X represents motion of PU in space. Axis I represents level of Cross-Shield interaction.

b. X_0-X_1 is the location where zero level of Cross-Shield interaction exists.

c. X_1-X_2 (A_1' is the location where Cross-Shield interaction changes its value from zero level to full interaction. (I_f is level of full interaction.) The changing value of Cross-Shield interaction is the main characteristic of that location.

65

d. X_2-X_3 (B_1) is the location where the level of Cross-Shield interaction reaches its highest value (I_f). This is a very special location where Cross-Shield interaction and Side-Shield interaction are equal to each other and are also equal to the maximum level of interaction. Side-Shield interaction means interaction between PU located on the same side of E-Shield.

e. X_3-X_4 (A1) is the location with characteristics equal to X_1-X_2. That location has the same mark (A1).

f. X_4-X_5 (C) is the location where zero level of Cross-Shield interaction exists. That is the "zoon" of E-Shield itself. On that zoon the E-Shield is fully established and keeps its usual characteristics (zero level of Cross-Shield interaction).

g. X_5-X_6 is the location with equal characteristics to X_1-X_2 and X_3-X_4 zones (A-type zones).

h. X_6-X_7 is the location with equal characteristics to X_2-X_3 zone (B-type zones).

i. X_7-X_8 is the location with equal characteristics to X_1-X_2 and X_3-X_4 zones (A-type zones).

9.3. 4. The most interesting locations mentioned in figure [9.3. 2.A] are B-type zones. Those locations have an equal level of Cross-Shield interaction and Side-Shield interaction. As a result any PU positioned on such a zone has full interaction between PU positioned on each side of E-Shield. Hence, CE-Space, HE-Space, and its own location look equal for such a unit.

9.3. 5. Equality between the levels of Cross-Shield interaction and Side-Shield interaction leads to the possibility that any PU has the same "behavior" (response on physical influence) according to its usual physical properties as it exists in undisturbed space.

9.3. 6. In the case of a moving object, it keeps the motion on B-type zoon as well as at any other places [b. 9.3. 5]. So B-type zones are only placing where the E-Shied can be penetrated by PU (objects, fields, etc.).

9.3. 7. In the three-dimensional world, A-type zones surround B-type zones entirely. As a result any B-type zone is reachable only by passing through an A-type zone.

9.3. 8. Therefore, B-type zones are only gaps that can be used to interact with physical units located in CE-Space and HE-Space.

9.3. 9. The following figure represents the rules of interaction between CE-Space and HE-Space. It shows two diagrams put one above other. The first diagram is the same one as [9.3. 2.A]. The second one is a diagram of a moving object's trajectory marked by numbers.

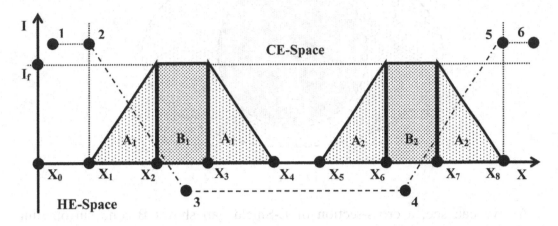

Fig. A

9.3. 10. Numbers on figure [9.3. 9.A] have the following meaning.

a. 1-2 is the part of trajectory on CE-Space. No disturbance exists on that part of trajectory. That is the first part of RW-Trajectory.

b. 2 is the point where Cross-Shield interaction begins to appear. It coincides to X_1 point of space.

c. 2-3 is the part of trajectory that leads to HE-Space. As mentioned before that part of trajectory is possible only in space enclosed by points X_2-X_3 (E-Shield gap).[1] Here and later I refer to that part of trajectory as Cross-Trajectory.

d. 3-4 is the part of trajectory located in HE-Space. That is Z-Trajectory itself.

e. 4-5 is the second part of Cross-Trajectory that leads an object through the next E-Shield gap (X_6-X_7) to CE-Space.

f. 5-6 is the part of trajectory in CE-Space. No disturbance exists on that part of trajectory. That is the next part of RW-Trajectory.

9.4 E-Shield Gap in Space

9.4. 1. Paragraph [9.3. 7] mentions the relation between B and A types of zones. A diagram of those zones and their cross-section (in the three-dimensional world) is shown in the next figure.

---

[1] In the figure.

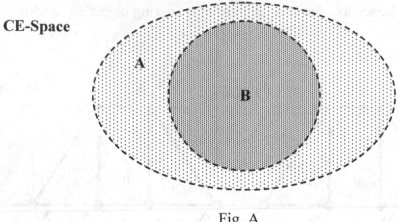

CE-Space

Fig. A

9.4. 2. As we can see, a cross-section of E-Shield gap shows B-zone surrounding completely by A-zone. In other words A-zone separates B-zone from CE-space and vice versa.

9.4. 3. As mentioned above, CE-Space has zero level of Cross-Shield interaction. In A-zone that interaction changes from zero to maximum level.[1] B-zone has a full level of Cross-Shield interaction.

9.4. 4. Here and later I refer to B-zone as Shield-Gap and A-zone as Gap-Hood. According to a moving object, we have two types of Shield-Gaps: those that are Out-Gap and those that are In-Gap. The first one means a Shield-Gap where an object goes from (out) CE-Space, and the second one means a Shield-Gap where an object goes to (in) CE-Space.

9.5 Shield-Gap and S-Outline

9.5. 1. It's time to look back to paragraph [7.1. 27] to refresh our memories about the notion of S-Outline and some other aspects mentioned in [9.3. 10].

9.5. 2. On the one hand, S-Outline permits a hidden motion of an object in a conservative filed. It's possible only between numbers of S-Outline points.

On the other hand, Z-Trajectory is possible only between two Shield-Gaps. Two those possibilities lead us to the following conclusion about gap properties of E-Shield.

9.5. 3. *Shield-Gaps can only exist at leading points of CE-Space that belong to S-Outline.*

---

[1] Some level of Cross-Shield interaction at each point of A-zone that lies between zero and maximum values.

9.5. 4. That is the only condition where conservative fields produce zero work and change no energy of an object moving in (to HE-Space) through one Shield-Gap and out (from HE-Space) through another one.

9.5. 5. The energy change of an object moving by Z-Trajectory between points 3 and 4 [s.9.3. 9.A] equals zero because of the zero level of Cross-Shied interaction despite the distance separating those two Shield-Gaps in CE-Space. That happens at Z-Trajectory in HE-Space as well as at any RWT in CE-Space using the same points as corresponding ones (belonging to same RW-Trajectory or Z-Trajectory).

9.5. 6. Therefore, relocation of an object through the trajectory mentioned as 2-3-4-5 [s.9.3. 9.A] changes no energy of an object and conservative fields, and keeps the law of conservation untouched.

9.5. 7. Any other way is impossible because it leads to a violation of the law of conservation.

Chapter 10. Matter of Time

10.1 Circular Trajectory

10.1. 1. Figure [5.1. 13.A] shows the number of possible Z-trajectories. All of them have first and last point located on S-Outline. In an ideal scenario S-Outline has the shape of an exact sphere around a single central body [7.1. 28].

10.1. 2. In case of an object moving in a circular orbit [s. 5.1. 13.A] around the central body it, always exists on the S-Outline. Therefore, it can be always be relocated by Z-Trajectory to any other point of its RWT around the central body. That relocation changes nothing in energy of the object and the conservative field, as mentioned earlier [r.4.2. 6].

10.1. 3. Moreover, the motion of an object by a circular trajectory around the central body changes no energy in the whole system at any time. As long as circular motion is undisturbed[1] by additional forces, an object keeps its circular trajectory and constant speed. In that case the law of conservation persists exactly (an object changes no potential or kinetics energy). As a result the object follows its trajectory circle by circle for any length of time.

10.1. 4. In such conditions, relocation of the object by Z-Trajectory S_1-S_2 (for example [s. 5.1. 13.A]) changes nothing in the energy of the whole system because sooner or later that object reaches S_2 point by its RWT. If the process of relocation by Z-Trajectory takes very little time, the object changes its location almost immediately[2] (unlike RWT).

10.1. 5. If the same process takes the same time as it needs to move the object by RWT from S_1 point to S_2 point, there is no difference in the state of the object. In other words the object takes point S_2 at the same time using RWT or Z-Trajectory.

10.1. 6. The more distant point used as the last point of Z-Trajectory, the more time needed for the object to reach that point by RWT. As a result the more time delay that appears in Z-Trajectory.

For example, the time delay of Z-Trajectory for S_1-S_3 transposition is greater than the time delay for S_1-S_2 transposition.

10.1. 7. Hence, when the time delay of Z-Trajectory reaches the period of revolution for an object around the central body, we have the same object at the same time and at the same place, regardless of using RWT or Z-Trajectory. As a result such displacement of the object exists only in the area of time.

---

[1] **And the whole system exists as isolated one.**

[2] The law of conservation restricts an object to use additional acceleration that way (that is not controversial).

10.1. 8. Moreover, using any number of revolutions, the object has the same result of displacement by Z-Trajectory and exists at the same point with displacement only in the area of time. The following equations show that mathematically.

$$D_z(t) = n \cdot T \qquad \text{(a)}$$

$$D_z(l) = 0 \qquad \text{(b)}$$

$D_z(t)$ is the displacement by Z-Trajectory in time, n is any integer number, T is the revolution time. $D_z(l)$ is the displacement by Z-Trajectory in space (length of trajectory) in a coordinate system bound to the central body.

10.1. 9. In this case the circular motion equation [10.1. 8. a] is not strong because any point of S-Outline lies on such a trajectory. Therefore, we can calculate the time lag between any two locations of a moving object. As a result the moving object can be displaced by Z-Trajectory for any point of its RWT. The equation shows that mathematically.

$$D_z(t) = R \qquad \text{(a)}$$

Where $D_z(t)$ is displacement by Z-Trajectory in time, and R is any real number.

10.1. 10. R and n in equations [10.1. 8. a] and [10.1. 9. a] can be positive or negative because a negative value means a previous location of the object.

10.1. 11. **Conclusion.** In the case of displacement of an object by Z-Trajectory when the displacement in time[1] is greater than zero, there is a time delay in the object's observation. During that time it looks invisible and unreachable[2] from CE-Space. Such a time delay can be as great as the period of existence of conservative field around the central body.

10.2 Elliptic Trajectory

10.2. 1. In the case of elliptic trajectory, we have the following figure that shows motion of an object around a central body. That is usually the state for celestial bodies that move around the stars because "the orbits of the planets are ellipses with the Sun at one focus".[3]

---

[1] $D_z(t)$
[2] Uninteractable, so to say.
[3] "Celestial mechanics." Encyclopaedia Britannica. <u>Encyclopaedia Britannica 2008 Deluxe Edition</u>. Chicago: Encyclopaedia Britannica, 2008.

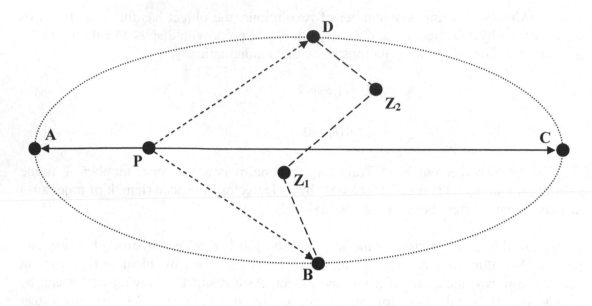

Fig. A

In figure [A]:
- P is the central body
- A-B-C-D-A is an elliptic trajectory
- P occupies one of the ellipse focuses
- B-Z_1-Z_2-D is an example of Z-Trajectory

10.2. 2. One aspect of elliptic trajectory (more important according to Z-Trajectory) is the variable distance between the object and the central body around which it moves. As you can see in figure [10.2. 1.A], point A is the closest location to the central body (point P) and point C is farthest one. In celestial mechanics, those points have special names: perihelion and aphelion.

10.2. 3. "The Sun occupies one of the two foci of the ellipse of a planet's orbit. A line drawn through the point of the planet's closest approach to the Sun (perihelion) and farthest retreat (aphelion) passes through the Sun and is called the line of apsides or major axis of the orbit; one-half this line's length is the semimajor axis, equivalent to the planet's mean distance from the Sun."[1]

10.2. 4. For any object moving in an elliptic trajectory, the distance between the object and the central body varies from perihelion distance (minimal distance or length of PA [s. 10.2. 1.A]) and aphelion distance (maximal distance or length of PC [s. 10.2. 1.A]).

---

[1] "Orbit." Encyclopaedia Britannica. Encyclopaedia Britannica 2008 Deluxe Edition. Chicago: Encyclopaedia Britannica, 2008.

10.2. 5. During one full circle of motion, the body once reaches a maximum and minimum distance from the central body and twice reaches any value of the other distances. Figure [s. 10.2. 1.A] shows such conditions of moving object by points A (point of minimal distance), C (point of maximal distance), and the B and D points of some equal distance (PB=PD).

10.2. 6. As shown in [3.2. 10], two or more points equidistant from the central body have an equal value of S_g (SGF). Equation [5.1. 10. A] shows the basic law for first and last point of Z-Trajectory. Applying it to figure [s. 10.2. 1.A], we have following.

$$S_b(x_1, y_1, z_1, t_1) = S_d(x_2, y_2, z_2, t_2)$$ (a)

Where S_b is conservative field strength at point B, and S_d is conservative field strength at point D.

10.2. 7. Using the same method, we have the following equation for points A and C (and B for an example).

$$S_a(x_a, y_a, z_a, t_n) \neq S_c(x_c, y_c, z_c, t_n) \neq S_b(x_b, y_b, z_b, t_n)$$ (a)

S_a, S_b, and S_c have same meaning as in equation [10.2. 6.A] for appropriate points. Therefore, there is only one possible case for the presence of Z-Trajectory. That case exists when an object relocates from point B to point D and vice versa. That is the only way to follow the rule for first and last point of Z-Trajectory [s. 5.1. 10–5.1. 11]. As a result we have can conclude the following.

10.2. 8. *In cases of elliptical trajectory, the surrounding central body of each S-Outline shrinks for two points equidistant from the central body and lies on the trajectory.*

10.2. 9. Such an S-Outline creates the possibility for an object to use Z-Trajectory almost immediately when time or relocation has little difference from zero. Unlike the motion of an object by an elliptic trajectory, this happens because an object keeps the same value of energy at both points. As a result it does not need not to wait to change energy in the usual way that happens at RWT.[1] In other words, the energy level of an object keeps constant at first and last point of Z-Trajectory. The next equation shows that type of relocation mathematically.

$$S_b(x_b, y_b, z_b, t_1) = S_d(x_d, y_d, z_d, t_1)$$ (a)

As we can see, the moment of time on both sides of equation A is identical to each other (t_1). Only the coordinates of relocating object have changed.

---

[1] In the case of elliptic trajectory, energy of the object changes from potential to kinetic and vice versa. It needs some time to change any amount of energy (doesn't matter how little it is) to a different type of energy.

Here and later I refer to such type of relocation by Z-Trajectory as Zero Time Relocation (ZTR).

10.2. 10. There is a significant difference for points A and C. Because those points are not equidistant from central body P [s.10.2. 1. A], there is no possibility for an object to have Z-Trajectory between them. But it is quite possible to have Z-Trajectory between those points after one or more circles of a RWT moving object.

10.2. 11. Such a case is shown at [10.1. 8] for circle trajectory. In the case of elliptical trajectory, only points A and C match equations [10.1. 8.a] and [10.1. 8.b] strongly. As a result [10.1. 7] is fully applicable for such points.

10.2. 12. Equation [5.1. 10. a] transfers for points of perihelion and aphelion to the following equation.

$$S_a(x_a, y_a, z_a, t_1) = S_a(x_a, y_a, z_a, t_2) \qquad \text{(a)}$$

Time point t_1 and t_2 mentioned in the equation above coincide to moments when the object reaches point A,[1] and they have following relation.

$$t_2 = t_1 + n \cdot T \qquad \text{(b)}$$

Variables n and T have the same meaning as for [10.1. 8.a]. In that case first and last point of Z-Trajectory coincides to each other only in space (not in time).

10.2. 13. Here and later I refer to such relocation by Z-Trajectory as Zero Length Relocation (ZLR).

10.2. 14. In general cases for elliptical trajectory, we have the following equation for time difference between first and last point of Z-Trajectory.

$$t_2 = t_1 + n \cdot T \pm \tau \qquad \text{(a)}$$

$$\tau \in [0; T/2] \qquad \text{(b)}$$

In equations A and B, t_1 is the moment of time when the moving object reaches perihelion, n and T have same meaning as for [10.1. 8.a], τ is period of time (for first point of ZT) that is more than zero [s.10.2. 14.b] and less than half of revolution period, and t_2 is the seeking point of time for last point of Z-Trajectory.

10.2. 15. Equation [10.2. 14.a] gives us two points in space that are positioned so that line AC [s.10.2. 1.A] mirrors those two points. Each pair of those points is equidistant from the central body with equal S_g and are the only possible location for first and last points of Z-Trajectory.

---

[1] Point C has the same law.

74

10.2. 16. Hence, unlike circular trajectory, elliptical trajectory permits only two points where the first point of ZT and the last point of ZT [s.9.4. 4] coincide with each other (points of perihelion and aphelion), and the number of pairs where the major axis of the orbit [s.10.2. 3] mirrors the points of RWT becomes the possible first and last points of ZT.

10.2. 17. Here and later I refer to those points as Z-Trajectory Head Points (ZTHP or HP). Other points of that trajectory become Z-Trajectory Tail Points (ZTTP or TP).

Chapter 11. Z-dynamics

11.1 Orbital Relation

11.1. 1. Previous chapters show some aspects of Z-Trajectory and relating things in more or less static circumstances. This part discusses the dynamic aspects of the same things. The following figure shows a diagram of two bodies relative to locations according to aspects of Z-Trajectory.

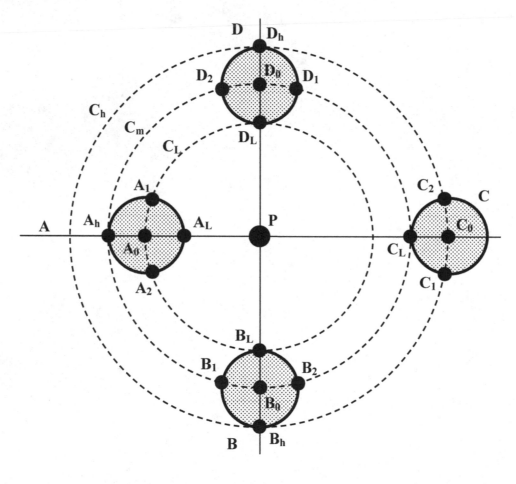

Fig. A

In figure A, point P is the location of the central body with a mass much more than the mass of the body (MB, moving body) moving by an elliptic trajectory A_0-B_0-C_0-D_0-A_0. Those points represent a few locations of the moving body center.

There are three concentric circles C_h, C_m, and C_L, whose centers coincide with point P. Circle C_h has the highest radius, circle C_L has the lowest radius, and circle C_m (middle circle) represents any circle with radius more than the radius of C_L and less than the radius of C_h. Each circle represents a number of points with equal SGF produced by the central body.

At each location (A_0-B_0-C_0-D_0) there are a few significant points mentioned as subscripts: 1, 2, L, and H. For example, there are four such points (B_L, B_h, B_1, B_2) at point B_0. Subscript L means the lowest point around an appropriate location that has the closest location to point P (lowest position accordingly central body). Subscript H means the highest point around an appropriate location that has the most distanced location from point P (highest position accordingly central body). Subscript 1 and 2 mean a number of points located on circle C_m and equidistant from the appropriate location (A_0, B_0, C_0, or D_0).

According to the distance between point P and locations (A_0, B_0, C_0, D_0), we have the following relations: $PA_0 < PB_0$; $PA_0 < PC_0$; $PA_0 < PD_0$; $PB_0 = PD_0$; $PD_0 < PC_0$. Accordingly, at other points we have the following relations: $A_0A_1 = A_0A_2 = A_0A_L = A_0A_h = B_0B_1 = B_0B_2 = B_0B_L = B_0B_h = C_0C_1 = C_0C_2 = C_0C_L = C_0C_h = D_0D_1 = D_0D_2 = D_0D_L = D_0D_h = R$, where R is static radius. It represents the radius of a sphere that covers the central point of a moving body in the three-dimensional world. In figure A it represents the radius of a circle drawn around the central point of a moving body (points A_0-B_0-C_0-D_0). In the case of a spherical body, R represent the radius of the spherical surface of the moving body itself.

11.1. 2. Hence, points 1, 2, L, and H around each location, mentioned as A_0, B_0. C_0, D_0, lie on an exact circle that has equal SGF of the moving body at each position.

11.1. 3. Because a section of a sphere touching another sphere (on their surfaces) in the three-dimensional world is a circle, a straight line positioned between points with subscript 1 and 2 represent the projection of those circles on a figure's surface (not drawn on the figure).

11.1. 4. The SGF at each point is the sum of SGF produced on the same point by the gravity field of a central body and a moving one. Therefore, according to the general law of Z-Trajectory [s.10.2. 14–10.2. 17], Head Points of that trajectory are able to exist only on the points of space where SGF are equal to each other (S-Outline definition [s.7.1. 27]).

11.1. 5. At the time when a moving body reaches point A_0, it has its closest location to the central one. That position allows S-Outline to exist as an A_1-A_2 circle.[1] As a result S-Outline permits the existence of HPZT at an exact circle. Hence, such an S-Outline permits Zero Time relocation between any pair of its points.

11.1. 6. In the case of reducing the radius of C_L, circle A_1-A_2 shrinks gradually to A_L point. Hence, the S-Outline shrinks to a single point (A_L) as well. It such a case there is a possibility only of Zero Length relocation by Z-Trajectory, because A_L point becomes

---

[1] That circle is mentioned in the figure as the most distant points (from each other) of that circle projection (A_1-A_2).

the only Head Point of Z-Trajectory. Hence, that point can be reached only at the time when the moving body is located at point A_0. Therefore, Z-Trajectory becomes possible only after n time of the moving body's revolution periods. At the time when n is multiplied with T (period of a full revolution [r.10.2. 12. B]), the moving body takes the same point (A_0), and $S_g(A_L)$ becomes exactly equal to the value of the previous circle. That can be shown mathematically by the following equation.

$$S_g(A_L, t_1) = S_g(A_L, t_1 + n \cdot T) \qquad \text{(a)}$$

In equation A, t_1, n, and T have same the meaning as for [10.2. 14. A]

11.1. 7. In the case of a rising radius of circle C_L, we have the same condition. When the radius of C_L becomes equal to the radius of C_m, the S-Outline shrinks from circle A_1-A_2 to A_h point. As a result aspects of the A_h point exactly match the same aspects of A_L point because those points become the only possible points of S-Outline if a moving body positioned itself at A_0 point.

11.1. 8. Moreover, A_h and A_L points cannot be HPZT to each other because of the S_g difference at those points at the time when the moving body takes position at A_0 and at any other point of time. The following equations show that mathematically.

$$S_g(A_L, t_n) \neq S_g(A_h, t_n) \qquad \text{(a)}$$

$$S_g(A_L, t_n) \neq S_g(A_h, t_m) \qquad \text{(b)}$$

Putting those equations in words, we have the following statement. The strength of conservative field at points A_h and A_L are always different (equation B) despite the moving body location and the number of its revolution around central body.

Equation A means a constant difference between the S_g of points A_h and A_L at any given point of time (different space points at same time).

11.1. 9. After some time the body moving in an elliptical trajectory reaches point B_0. As mentioned above, $PA_0 < PB_0$. In such a position, the moving body has an S-Outline as circle B_1-B_2 that shrinks to point B_L as the closest point to central body and B_h as the most distant one.

11.1. 10. Despite equal value of S_g of the central body and the moving one, any points A_h and B_1 (for example) with full S_g on those points are quite different. Such a difference appears because of different angles between the vectors of gravity attraction of point A_h and B_1 (magnitudes of those vectors are equal to each other). As a result the full S_g becomes different. Therefore, those points are unable to be HPZT to each other, and Z-Trajectory is unable to exist between them.

11.1. 11. The same difference exists between points around position B_0 and C_0 because of the difference between PB_0 and PC_0 ($PB_0 < PC_0$). As a result $S_g(B_0) > S_g(C_0)$ because the less distance between the point and the central body, the more SGF produces on the body at that point.

11.1. 12. At location D_0 the SGF at specific points is equal according to the SGF of a point at B_0 location because of the equality of PD_0 and PB_0 ($PD_0 = PB_0$). Hence, the S-Outline D_1-D_2 becomes equal to the S-Outline B_1-B_2 (SGF at each point of circle D_1-D_2 equal SGF at each point of circle B_1-B_2). Here we have same condition as in the case of two equal bodies [s. 7.1]. As a result the S-Outline for such case includes two circles, B_1-B_2 and D_1-D_2.

11.1. 13. Locations B_0 and D_0 are the only ones that can possibly have an S-Outline not only around the moving body and around another location of that body. As a result the HPZT are able to exist around both locations of the moving body (B_0 and D_0). But the moving body reaches location B_0 at different times than D_0 because of elliptical trajectory (RWT for moving body). Hence, locations B_0 and D_0 are separated by the following time

$$t_{B_0} = t_{D_0} + 2 \cdot \tau_p \qquad (a)$$

Where t_{B0} is the time when the moving body reaches B_0 location, t_{D0} is the time when the moving body reaches D_0 location, and τ is the period of time the moving body travels from location D_0 to A_0. In other words τ_p is the period of time when the body reaches perihelion of its orbit moving from a specific location.

11.1. 14. Hence, any object that use Z-Trajectory by S-Outline from D_1-D_2 to B_1-B_2 must keep that trajectory until the moving body reaches location B_0(period of time 2τ). What's more, such a relocation for the object is not a symmetric one because of significant differences between the periods of time spent moving the body to reach location B_0 from location D_0 and from B_0 to D_0. That happens because of the elliptical trajectory the moving body keeps.[1]

11.1. 15. In the second case an object must keep Z-Trajectory until the moving body reaches location D_0 moving from B_0. That time can be calculated by following equation.

$$t_{D_0} = t_{B_0} + 2 \cdot \tau_a \qquad (a)$$

Where t_{B0} and t_{D0} have same meaning as for equation [11.1. 13.a], and τ_a is the period of time when moving body reaches aphelion (C_0) from point B_0 (point of RWT between perihelion and aphelion).

---

[1] That is the case of planetary motion by elliptical trajectory.

79

11.1. 16. In general cases $\tau_a \neq \tau_p$. In each orbit (circular or elliptic) there are only two points with equal value for τ_a and τ_p. Those points are separated by a time equal to half of the full period of the moving body's revolution (T / 2). At those two points there is no time difference between time delays for Z-Trajectory relative to moving body RWT, including perihelion or aphelion.

Here and later I refer to those time delays as perihelion time delay (PT-Delay) and aphelion time delay (AT-Delay). At specific points where PT-Delay equals to AT-Delay, there is a half-period delay (HT-Delay, half time delay, time meaning time of revolution period).

11.1. 17. Some time passes (τ_a) and the moving body reaches location C_0. The relation between that location and the previous one (B_0) is equal as in the relation between locations B_0 and A_0. Those locations have different distances from the central body. As a result S_g at each point (A_0, B_0, and C_0) has its own value (unique value). Therefore, any specific point of C_0 location cannot be HPZT for appropriate points of B_0 location.

11.1. 18. Locations C_0 and A_0 have the same difference. As a result Z-Trajectory is impossible between specific points of C_0 and A_0.

11.1. 19. All reasoning mentioned above in this part of the work leads us to the following deduction in cases of a moving body's elliptical trajectory.

  a. *Z-Trajectory is possible between any points of S-Outline at any location of the moving body that are zero distance relocation.* In that case Z-Trajectory leads from any point of S-Outline to any other point of the same S-Outline when the moving body takes exactly the same location relative to the central body within any number of full time revolutions.
  b. *Z-Trajectory is possible between any points of S-Outline at any location of the moving body as zero time relocation.* In that case Z-Trajectory leads from any point of S-Outline to any other point of the same S-Outline while the moving body takes any exact location. Such relocation is possible around the moving body (for example, the circles mentioned by subscript 1 and 2, A_1-A_2, B_1-B_2, etc.).
  c. *Z-Trajectory is possible between any points of S-Outline at the location of the moving body that is mirrored by a major axis of the orbit* (A_0-C_0 on fig. [11.1. 1.A]). In that case the time delay for an object using Z-Trajectory is equal to double the time of the approaching perihelion or aphelion (according to the closest location of the moving body).
  d. *Z-Trajectory is* impossible *between points of S-Outline at the moving body's at perihelion and aphelion* (A_0-C_0 on [11.1. 1.A]). In that case time delay for an object using Z-Trajectory equals infinite because there is no time point where S_g at point A_0 equals to S_g at point C_0 [s.11.1. 1.A].

11.1. 20. As we can see in the case of a moving body, it plays a key role in guiding ZTHP to exist at some particular points of space. Without a moving body's conservative field, the surrounding central body helps S-Outline exist as a number of concentric, undisturbed spheres around it.

11.1. 21. A body moving in a circular orbit always exists only on one S-Outline. Unlike a circular orbit, an elliptical one leads the moving body to move through a number of spherical S-Outlines of the central body. As a result ZTHP is possible to exist only at some points that have equal S_g.

11.2 Displacement Relation

11.2. 1. Previous sections discuss the possibility and properties of Z-Trajectory relative to the orbital motion of a body. This section discusses the same problem relative to Cross-Trajectory [9.3. 10.c] that leads an object to and from HE-Space. The following figure shows that process in space and time ordinates.

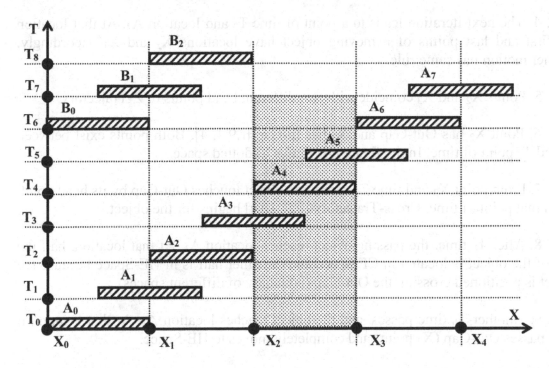

Fig. A

There are two axes in figure [A]. Axis X shows the relocation of objects in space, and axis T shows the relocation of objects by time. A and B are two objects. Each of them are shown as a number of locations related to both axes by number subscripts.

For example, A_0 means the location of object A at the moment of time T_0. At that moment it is located between points X_0 and X_1. Obviously, the difference between X_1 and X_0 gives us the length of the moving object ($L = X_1 - X_0$).

81

In the figure, $X_4-X_3 = X_3-X_2 = X_2-X_1 = X_1-X_0 = L$; $T_n-T_{n-1} = T_h$. The value of T_h means some period of time. According to figure [A], time T_h equals the period of time when the moving object covers a distance equal to half of its full length (L). Additional value T has the following relation to T_h ($T=2 \cdot T_h$). T is the period of time when a moving object covers the distance equal to its full length (L).

According to the figure, $(T_n - T_{n-1}) / (X_m-X_{m-1})$ = constant. That means a constant speed of the objects A and B because of their shown locations.

11.2. 2. At the beginning (T_0 moment of time), object A is located at its "head" point, positioned at X_1, and at its "tail" point positioned at X_0. The direction of an object's movement coincides with a positive direction of the X axis.

11.2. 3. After T_h (period of time), time itself reaches value T_1, and the object reaches location A_1, with its middle point positioned at X_1.

11.2. 4. The next iteration leads to a point of time T_2 and location A_2. At that location the first and last points of a moving object have locations X_2 and X_1 accordingly. Further motion has same rule.

11.2. 5. Points X_2 and X_3 coincide with two different head points of Z-Trajectory.

11.2. 6. Point X_2 has Out-Gap and X_3 has In-Gap [r.9.4. 4]. Both points exist between T_0 and T_7 point of time. In the figure it is drawn as dotted space.

11.2. 7. From point X_2 and time T_2, a moving object touches Out-Gap by its head point. From that point of time, Cross-Trajectory [9.3. 10.c] begins for the object.

11.2. 8. After T_h time, the passing object reaches location A_3. At that location, half the part of the object is located in CE-Space and the other half is in HE-Space because the object is positioned crossing the Out-Gap (boundary of different spaces).

11.2. 9. Another T_h time passes, and the object reaches location A_4. At that location it fully passes Out-Gap (X_2 point) and completely moves to HE-Space.

11.2. 10. All number of locations between A_2 and A_4 have special characteristics as part of the moving object is located in both spaces (CE and HE). That leads us to the following question: what happened if Out-Gap closed at any moment when only part of the object had passed though it?

11.2. 11. As mentioned in [8.5. 1.b], closing the Out-Gap mustn't violate the law of conservation.[1] Therefore, the sum of energy must be keep constant with the presence of Out-Gap full time or part time. Full time presence of Out-Gap means its presence for time enough for a moving object to pass the Out-Gap completely. Part time presence permits only some part of the object to pass through Out-Gap. In other words, part time

---

[1] As well as any other process in our world (CE-Space).

presence leads to the existence of Out-Gap only for a time that is less than one needs for the object to fully pass through the Gap.

11.2. 12. To adhere to the law of conservation, Out-Gap mustn't do anything to the object moving through it. If that gap changes, any physical characteristics of the moving object must use more or less energy for such a changing. That energy must be brought from the outside of CE-Space and would violate law of conservation. Therefore, that way is impossible because of the law of conservation's restriction.

11.2. 13. For example, Out-Gap cannot further "push" (change velocity by positive acceleration) the object to ensure it passes through before the gap is closed. Also, Out-Gap can not "pull" (change velocity by negative acceleration) the object to ensure it is moving back (keep on CE-Space) before the gap is closed.

11.2. 14. As a result there is only one way to keep the law of conservation for Out-Gap. It must leave the object before Z-Trajectory in case of closing the Out-Gap before the object passes the gap completely. In other words, the object is unable to start Z-Trajectory (change RWT to ZT) before it goes through Out-Gap with its full length.

11.2. 15. In case of closing Out-Gap at point X_2 [s. 11.2. 1.A] before T_4 point of time, the object A continues to keep RWT in CE-Space without any changes of its state and energy. Therefore, the presence or absence Out-Gap is significant for a moving object only when the object goes completely through the gap at the period of time when the gap exists. Here and later I refer to that as the law of Cross-Trajectory.

11.2. 16. In the case of X_3, point we have the same situation as for X_2 point. The difference between points X_3 and X_2 is that X_2 point represents Out-Gap and X_3 represents In-Gap. All reasoning mentioned at [11.2. 7–11.2. 15] is fully applicable to X_3 point with only one difference. Out-Gap leads the moving object back to CE-Space. As a result, the law of Cross-Trajectory is fully applicable to Out-Gap as well.

11.2. 17. In cases when a moving object goes through Out-Gap, it must reach at least location A_6 to pass Out-Gap completely [s. 11.2. 1.A]. At that location the object is located between X_4 and X_3 points. Otherwise the object keeps Z-Trajectory to adhere to the law of conservation.

11.2. 18. This leads us to the following question. Why is the gap unable to split the object passing partly through it in the case of a closing gap? That is easily explained: the splitting of any object or system that has any physical relation between their units needs more or less energy. Even the process of brooking the thinnest glass spends more or less energy. If that process appears at a gap, it violates the law of conservation [r.8.5. 1.b], which is impossible [r.8.3. 4].

11.2. 19. Accordingly [s. 11.2. 1.A], RWR trajectory for a moving system appears from location A_0 to A_2 and from A_6 to A_7. Z-Trajectory appears from locations A_2 to A_6.

Those locations (A_2-A_6) have a linear difference equal to X_4-X_2 (locations of head point). The equation shows that mathematically.

$$\Delta X = X_4 - X_2 = (X_4 - X_3) + (X_3 - X_2) = L + L = 2 \cdot L \qquad \text{(a)}$$

Where ΔX is displacement for moving system that goes to and from HE-Space by Z-Trajectory. Because X_n-X_{n-1}=L [r. 11.2. 1], then X_4-X_3=X_3-X_2=L as well.

11.2. 20. Therefore, the minimal distance that must be coved by relocation of a moving system going by Z-Trajectory between two different head points of Z-Trajectory must be equal to double the maximal length of the moving system (relative to the direction of relocation). Maximal length means the maximal distance between the two most remote points in the direction of moving.

11.2. 21. According to the length of displacement by Z-Trajectory (X_2-X_4 at [11.2. 1.A]), it is possible to calculate the relocation time that the moving system spends by going between two different head points of Z-Trajectory. The following equations show the calculations mathematically [r.4.1. 1. A].

$$V(t) = \frac{dX}{dt} \qquad \text{(a)}$$

In the case of a constant velocity, the equation turns to following one: $V = \Delta X / \Delta t$

$$T_Z = \frac{\Delta X}{V} = \frac{2 \cdot L}{V} \qquad \text{(b)}$$

In equation b the variable T_z is time of relocation, L is maximal length of moving system (in direction of displacement), and V is velocity of moving system.

11.2. 22. Here and further I refer to value T_z [s.11.2. 21.b] as <u>ZT-Time</u> (Z-Transposition Time). ZT-Time means the minimal period of time that must be spent by moving a system that uses Z-Trajectory to its full relocation from one head point of Z-Trajectory to the next one.

11.2. 23. It's time to turn our attention to the next moving system. System B in figure [11.2. 1.A] reaches point X_2 at T_8 point of time. Because a gap exists at that point only to T_7 point of time, system B reaches X_2 point at (T_8-T_7) time after it closes. As a result system B keeps RWT after X_2 point of space as well as before that point and uses no Z-Trajectory. Time delay of (T_8-T_7) (in the case of [11.2. 1.A]) prevents system B from being involved in the Out-Gap passing process.

11.2. 24. In following explanations I refer to the process of passing Z-Trajectory and its corresponding Shield-Gaps (Out-Gap and In-Gap) by any physical system as Z-Process.

Z-Theory Applications

Chapter 12. Physical Phenomena

12.1 "Cloud" Mistake

12.1. 1. Chapter 9 (especially [9.2]) discussed different levels of interaction that occur between a system located in CE-Space and at some points close to HE-Space. It's time to discuss some phenomena that help us to understand more clearly the characteristics of reduced interaction. Generally I start to apply aspects of the previously explained theory to the real world.

12.1. 2. As mentioned in [9.4], there is only one space where reduced interaction takes place: Gap-Hood [s.9.4. 4]. What does it means for an IB [r.2.1. 3.a] that stays outside of Gap-Hood? The following observations are possible in that case.

12.1. 3. First of all we need to think about the usual way that humans have information from the surrounding world. In the case of full interaction between objects of this world and the human senses, we can imagine the usual sequences of such interaction. The following sequence shows a number of steps of interaction between the human senses (vision) and the observing object (a white sphere, for example).

    a. Light goes from the light source to the sphere
    b. Light reflects from the sphere
    c. Light reaches the IO [r.2.1. 3.b]
    d. IO has a visual image of the white sphere

12.1. 4. What happens in the case of reduced interaction between the object and an IO? In the real world (CE-Space) we have some examples of such a process. First of all we need to remember the usual processes that reduce interaction between an observing object and an IO.

12.1. 5. In the case of reducing interaction by sending only some light to the IO from the observing object, the IO can see only some reduced image that lacks some light (energy of visible electromagnetic waves).

12.1. 6. That happens in the real world each time any substance exists between the IO and the observing object. It can be a solid substance or a substance suspension in air. Hence, light waves traveling through such space partly reduces the light's energy by light absorption (or changing the direction of light waves by refraction) or substance suspension, partly changing their ways and avoiding the IO. As a result only one part of the reflecting light wave's power (traveling from the observing object) reaches the IO.

12.1. 7. Usually a substance that has a suspension in air is water, such as fog. "Fog is cloud of small water droplets near ground level and sufficiently dense to reduce horizontal visibility to less than 1,000 m (3,281 feet). The word fog also may refer to clouds of smoke particles, ice particles, or mixtures of these components. Under similar conditions, but with visibility greater than 1,000 m, the phenomenon is termed a mist or haze, depending on whether the obscurity is caused by water drops or solid particles."[1]

12.1. 8. There is a very interesting feature of human vision that leads humans to "detect" fog at any place where is no clear vision. As a result human senses "feel" fog at any place that has a reduced interaction with IO.

12.1. 9. Therefore, each zone that has the same characteristics (reduced interaction) must be seen by human vision as "fog". In cases when such an area has a given location in air, it must be mistaken as a "cloud" because there is no way to separate feelings from real clouds and the area with reduced interaction. Hence, Gap-Hood must be observed from the outside as a cloud. Obviously, it can be observed only in cases of a light source presence that has enough power to produce a visual sensation for IO at the point of observation.

12.1. 10. Here and later I refer to such an occurrence as fog effect.

12.2 Rainbow Effect

12.2. 1. The effect mentioned in the previous topic appears in cases when the whole electromagnet spectrum is affected equally by reduced interaction. There are some different consequences if the spectrum affected in a different way.

12.2. 2. Figure [9.2. 3. A] shows an example of reduced interaction on some particular area. It can be applied to show different levels of interaction between an IO and the observing object. I use the same schema here as for [12.1. 3] (IO and the white sphere). The following figure shows reduced interaction in the case of "unequally reduced interaction".

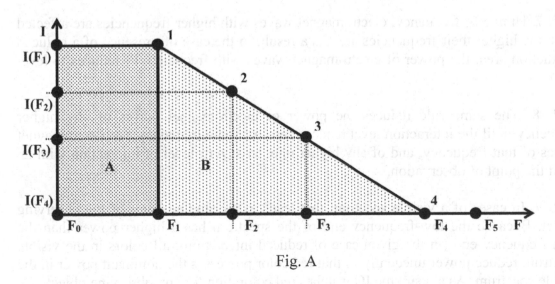

Fig. A

[1] "Fog." Encyclopaedia Britannica. Encyclopaedia Britannica 2008 Deluxe Edition. Chicago: Encyclopaedia Britannica, 2008.

In figure [A] there are two axes, F (frequency) and I (interaction level). The diagram itself shows the level of interaction between an IO located at some particular point and the spectrum of electromagnet waves at the same point of space. Bold lines (horizontal and angled one) on the diagram mean correspondence between frequencies and level of interaction. The same aspect affects sensations of IO.

F_n are different frequencies of electromagnet waves. $I(F_n)$ are different levels of interaction at the point of observation between the IO and light waves. Zone A is the zone of full interaction. Zone B is the zone of unequally reduced interaction.

12.2. 3. As we can see any frequency between F_0 and F_1 has full interaction with IO because of the horizontal line of interaction level in the diagram [12.2. 2.A]. Those frequencies have a level of interaction $I(F_1)$ that is equal to the interaction level of frequency F_1, where "unequally reduced interaction" begins (point 1).

12.2. 4. Frequency F_2 has a level of interaction $I(F_2)$ according to point 2. Frequency F_3 has a level of interaction $I(F_3)$ according to point 3, and frequency F_4 has a level of interaction $I(F_4)$ according to point 4. Frequency F_4 has zero level of interaction $I(F_4)$ because point 4 coincides with the axis of frequency. As a result $I(F_4) = 0$.

12.2. 5. According to figure [12.2. 2.A], the more subscript number of frequency, the higher the frequency itself ($F_0 < F_1 < F_2 < F_3 < F_4$). Unlike frequencies, levels of interaction have an inverse relation to the subscripts. Hence, $I(F_1) > I(F_2) > I(F_3) > I(F_4)$.

12.2. 6. In the case shown at figure [12.2. 2.A], electromagnet waves that have the same power initially reduce their power according to their frequencies, going through the area with reduced interaction. For example, electromagnet waves with frequency F_1 have equal power at the point of observation with or without Gap-Hood.

12.2. 7. Unlike F_1 frequency, electromagnet waves with higher frequencies are affected more the higher their frequencies are. As a result, in the case of presence of a reduced interaction area, the power of electromagnet waves with frequency F_2 reduces to level $I(F_2)$.

12.2. 8. The same rule reduces the power of electromagnet waves of any higher frequency until the interaction level reaches zero for some frequency F_4. Electromagnet waves of that frequency, and of any higher one, have zero level of interaction with an IO at the point of observation.

12.2. 9. In cases of a visible spectrum that leads to a changing color of the observing object. Because the low-frequency end of the spectrum has a higher power than the high-frequency end (in the given case of reduced interaction), all colors in the visible spectrum reduce power unequally so that red color possesses the dominant power in the visible spectrum. As a result the IO watches red coloration for any observing object.

12.2. 10. Generally any object that has coloration with more than one spectrum color (one color frequency) changes coloration so that low-frequency parts of coloration become more noticeable. For a white object or something with the same color, the IO watches red color coloration. Of course the coloration changing has no relation to the nature of the observing object or thing because that effect is caused only by an area of reduced interaction, not the object or thing itself.

12.2. 11. Here and later I refer to the effect of changing color of light caused by its passage through an area of reduced interaction as Rainbow Effect.

12.2. 12. Generally there are two possible ways for a Rainbow Effect to occur. If the area of reduced interaction affects the high-frequency end of the electromagnet spectrum, the IO watches a red coloration of surrounding space and objects. Otherwise the low-frequency end of the electromagnet spectrum is affected by an area of reduced interaction. In such cases IO watches a blue coloration of surrounding space and objects because the blue end of the visible spectrum obtains more power in such conditions of space.

12.2. 13. The following figure shows a case of blue end electromagnet spectrum with its higher power. The diagram looks similar to the case shown in the previous figure [12.2. 2.A] with only difference: the line of interaction level has a different angle (drawn as a broken one).

Points F_1, 5, 6, and 7 are points of interaction level. As we can see the higher the frequency, the less it is affected by the area of reduced interaction in cases when the low-frequency end of electromagnet spectrum is affected by an area of reduced interaction.

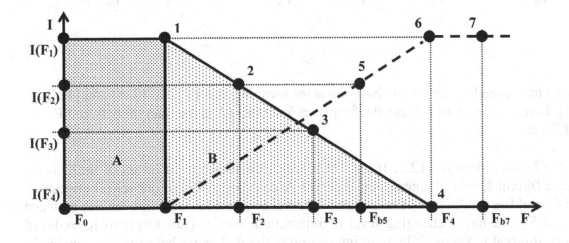

Fig. A

12.2. 14. Here and later I refer to the effect of high-end spectrum reducing interaction with IO as high shift on (of) the Rainbow Effect (HSRE). Otherwise there is a low shift on (of) the Rainbow Effect (LSRE) in cases of the low-end spectrum reducing interaction with IO.

In the visible spectrum HSRE leads to red coloration for units observed by an IO located at a point of reduced interaction area. Otherwise the IO watches blue coloration (LSRE) of the same objects at the same location.

12.2. 15. Figure [12.2. 13.A] clearly shows this. The level of reducing interaction $I(F_2)$ is reachable in the case of HSRE at frequency F_2 and in the case of LSRE at frequency F_{b5}. In other words, in case HSRE and LSRE are at the same level of reducing interaction but reach for different frequencies. That is the cause of different coloration from the perspective of IO at the point of observation in cases of HSRE or LSRE.

12.2. 16. Any other type of Rainbow Effect can be reduced to a combination of low and high shifts. An example is shown in the following figure.

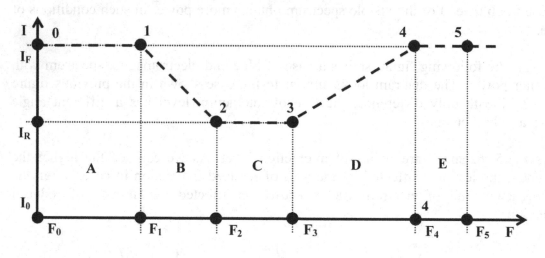

Fig. A

Marks mentioned on figure [A] have the same meaning as for [12.2. 13.A] and [12.2. 2.A]. Letters from A to E mean the frequency intervals (A is the interval between F_0 and F_1, etc.).

12.2. 17. According to [12.2. 16.A], there are a few different intervals of frequencies with different levels of interaction. Interval A coincides with frequencies between F_0 and F_1 and has a full level of interaction. Interval B coincides with frequencies between F_1 and F_2 and has a changing level of interaction from I_F (full level) to I_R (reduced level). Interval C keeps its level of interaction without changes between frequencies F_2 and F_3. Interval D looks like B with only one difference: the level of interaction on the frequency interval between F_3 and F_4 rises in value from I_R to I_F. Interval E has a full level of interaction as well as interval A.

12.2. 18. Interval B [s.12.2. 16.A] represents HSRE, Interval D represents LSRE. Interval C represents the fog effect that leads to a reduction of interaction equally in some frequency interval (between F_2 and F_3). As a result an IO watches some coloration with vision covered by some fog.

12.3 Radio Communication

12.3. 1. To show the relationship between the previous topic and the practical use of electromagnet waves, I mention below some basic principles of radio and its relation to electromagnet waves[1] according to modern science.

12.3. 2. "Radio is the transmission and detection of communication signals consisting of electromagnetic waves that travel through the air in a straight line or by reflection from the ionosphere or from a communications satellite."[2]

12.3. 3. "Electromagnetic radiation includes light as well as radio waves, and the two have many properties in common. Both are propagated through space in approximately straight lines at a velocity of about 300,000,000 metres (186,000 miles) per second and have amplitudes that vary cyclically with time; that is, they oscillate from zero amplitude to a maximum and back again. The number of times the cycle is repeated in one second is called the frequency (symbolized as f) in cycles per second, and the time taken to complete one cycle is 1 / f seconds, sometimes called the period. To commemorate the German pioneer Heinrich Hertz, who carried out some of the early radio experiments, the cycle per second is now called a hertz so that a frequency of one cycle per second is written as one hertz (abbreviated Hz)."[3]

12.3. 4. "A radio wave being propagated through space will at any given instant have an amplitude variation along its direction of travel similar to that of its time variation, much like a wave traveling on a body of water. The distance from one wave crest to the next is known as the wavelength."

12.3. 5. "Wavelength and frequency are related. Dividing the speed of the electromagnetic wave (c) by the wavelength (designated by the Greek letter lambda, λ) gives the frequency: $f = c / \lambda$. Thus a wavelength of 10 metres has a frequency of 300,000,000 divided by 10, or 30,000,000 hertz (30 megahertz). The wavelength of light is much shorter than that of a radio wave. At the centre of the light spectrum the wavelength is about 0.5 micron (0.0000005 metre), or a frequency of 6×10^{14} hertz or 600,000 gigahertz (one gigahertz equals 1,000,000,000 hertz). The maximum frequency in the radio spectrum is usually taken to be about 45 gigahertz, corresponding to a

---

1 Mostly for the readers who are not quite familiar with that area of physics.
[2] "Radio." Encyclopaedia Britannica. Encyclopaedia Britannica 2008 Deluxe Edition. Chicago: Encyclopaedia Britannica, 2008.
[3] "Radio." Encyclopaedia Britannica. Encyclopaedia Britannica 2008 Deluxe Edition. Chicago: Encyclopaedia Britannica, 2008.

wavelength of about 6.7 millimetres. Radio waves can be generated and used at frequencies lower than 10 kilohertz ($\lambda = 30,000$ metres)."[1]

12.3. 6. Paragraphs [12.3. 3–5] give us important information about the relationship between different frequencies of electromagnet waves and their appearance. Only a short part of the electromagnet spectrum is used by human vision. Other parts of the spectrum are used only by special devices (radio communicators, radars, etc.). Therefore, for any device that uses a frequency (or combination of frequencies), for it to work it must be affected by the Rainbow Effect regardless of its nature (human vision looks similar to a radio receiver that way). That is the main reason why I prefer not to separate types of electromagnet wave receivers and use the name of Independent Observers for all of them.

12.3. 7. In such a case, figure [12.3. 2.A] can be applied to different frequencies of visible light and radio waves (and to any other electromagnet waves). For example, if frequencies F_2-F_3 represent the visible band of electromagnet spectrum (F_2 in the red end of the spectrum and F_3 in the blue one) at a particular point of IO's location, then frequency F_1 represents the carrier frequency of some radio transceiver at the same point.

12.3. 8. Figure [12.3. 2.A] shows HSRE [s.12.2. 14]. As a result anything located at that area has a red coloration on the visible spectrum for a human observer. Unlike a person, a radio transceiver has (notices) no change in radio communication because its carrier frequency (F_1) is not affected by the Rainbow Effect. The following equation shows that mathematically.

$$I(F_c) = I(F_0) = \max \tag{a}$$

In the equation $I(F_c)$ is interaction level between radio transceiver and carrier frequency radio waves at the point of device location. It equals the maximal level of possible interaction between the transceiver and the radio waves.

12.3. 9. In such a case, the human would notice red coloration of the surrounding area and be able to send that information by radio transmission. That is the only reason for some evidences about a red sky that was received by the radio from some of the that pilots suddenly found them at an area with HSRE.

12.3. 10. It is the same reason (Rainbow Effect) that causes an absence of any evidence about other sky coloration. In the case of LSRE [s.12.2. 13.A], (broken line) carrier frequency (F_1) has zero level of interaction with a radio transceiver at the time then visible light waves have reduced interaction (frequency F_{b5}) and can be noticed by a human observer (IO).

---

[1] "Radio." Encyclopaedia Britannica. Encyclopaedia Britannica 2008 Deluxe Edition. Chicago: Encyclopaedia Britannica, 2008.

12.3. 11. As a result a human watches strange coloration of the surrounding space (green sky or blue sky) but is unable to send that information because the radio channel is unable to transfer radio signals at the carrier frequency (F_1) to any IB located outside of the area with LSRE. That is the reason for an absence of any evidence about other types of sky coloration received from any eyewitness by radio channel.

In such cases the level of interaction ($I(F_c)$) between radio waves from CE-Space and transceiver (and between radio waves from transceiver and CE-Space) can be shown by the following equation.

$$I(F_c) = 0 \tag{a}$$

12.3. 12. The notion "level of interaction between radio waves and transceiver" mentioned above means the difference of radio waves' power received by any transmitter from other transmitters (or sources of electromagnet waves), and vice versa in cases of the presence and absence of reduced interaction around the location point of the transceiver. If the area of reduced interaction is absent, the level of interaction always equals the maximum (relative to a given point of transmitter location).

12.4 Operation of Radars

12.4. 1. As is commonly known, there is one more special device that uses radio waves to operation. That device has name radar. This topic discusses aspects of radar operation with an object located in a reduced interaction area.

12.4. 2. "Radar is electromagnetic sensor used for detecting, locating, tracking, and recognizing objects of various kinds at considerable distances. It operates by transmitting electromagnetic energy toward objects, commonly referred to as targets, and observing the echoes returned from them. The targets may be aircraft, ships, spacecraft, automotive vehicles, and astronomical bodies, or even birds, insects, and rain. Besides determining the presence, location, and velocity of such objects, radar can sometimes obtain their size and shape as well. What distinguishes radar from optical and infrared sensing devices is its ability to detect faraway objects under adverse weather conditions and to determine their range, or distance, with precision.

12.4. 3. "Radar is an 'active' sensing device in that it has its own source of illumination (a transmitter) for locating targets. It typically operates in the microwave region of the electromagnetic spectrum – measured in hertz (cycles per second), at frequencies extending from about 400 megahertz (MHz) to 40 gigahertz (GHz). It has, however, been used at lower frequencies for long-range applications (frequencies as low as several megahertz, which is the HF [high-frequency], or shortwave, band) and at optical and infrared frequencies (those of laser radar, or lidar)."[1]

---

[1] "Radar." Encyclopaedia Britannica. Encyclopaedia Britannica 2008 Deluxe Edition. Chicago: Encyclopaedia Britannica, 2008.

12.4. 4. The most important characteristic of radars relative to this work is their sensitivity. Each radar receives an echo signal from anything that mirrors its signal and decides whether it is a target or not. Therefore, any radar has some minimal level of a mirroring signal that can be used in order to properly work. If the reflecting signal power falls below that level, radar is unable to detect a target. The following figure shows that graphically.

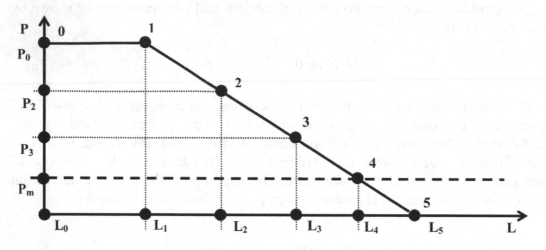

Fig. A

12.4. 5. Figure [12.4. 4.A] shows two axis, P and L. Those are the reflecting power signal, and the level of interaction between the target and electromagnet waves emitted by radar, respectively. Points mentioned by numbers represent different levels of interaction (the state of the area of reduced interaction at the point of the target location).

12.4. 6. Initially the target positioned at some point of space (0) that is located outside of the reduced interaction area. The reflecting signal from that point has power P_0. Then the target reaches some different point where a reduced area of interaction begins (1); the reducing signal power has no changes because the edge of ARI (area of reduced interaction) produces no changes to the reflecting signal.

12.4. 7. At point 2, ARI produces some influence and reduces the interaction between the target and radar signal so that the reflecting signal drops its power to P_2 level ($P_2<P_0$). It relates only to state of area surrounding the target, not to the distance between the radar and the target or atmospheric disturbance.

12.4. 8. The same process occurs until the target reaches point 4. At that point the interaction between the target and radar signal reduces dramatically, and the reflecting signal power reaches the minimal possible level for correct radar operation (P_m, minimal acceptable power of reflecting signal).

12.4. 9. Any location of the target between points 4 and 5 [s. 12.4. 4.A] is undetectable by the radar because those locations have very little level of interaction between the target and the radar signal. The equation shows that mathematically.

$$P_r < P_m \qquad \text{(a)}$$

Where P_r is power (at radar location) of signal reflected by target, and P_m is minimal echo signal power acceptable for radar correct operation.

12.4. 10. In cases when ARI is absent, each point (1–5) [s. 12.4. 4.A] has the same level of interaction between the radar signal and the target. As a result each point has an echo signal power equal to P_0.

12.4. 11. Therefore, the ARI presence leads to the radar's inability to locate (trace) any target positioned inside the area where equation [12.4. 9.A] is correct.

12.4. 12. Under usual circumstances there is no way to separate a reduced echo signal power from atmospheric influence (or target characteristics) and the presence of ARI, because at the point of radar location both processes have the same result (insufficient power of analyzing signal) from the radar's point of view. That is the reason why typical radar is unable to see not only the target inside ARI (or Gap-Hood) but the area itself (as well as Gap-Hood).

12.4. 13. In any case where ARI exists at any particular place, it interacts with electromagnet waves by the way explained at [12.3–12.4]. It is worth posing the following question: where is the additional amount of electromagnet energy that has no interaction with IO at the point of observation? That question can be easily answered.

12.4. 14. That energy has no interaction with the IO because it passes through the ARI without any change. That is the primary aspect of ARI. In other words, reduced interaction means reducing the electromagnet spectrum that interacts with the IO at any location inside such an area. The following picture shows that characteristics of describing area.

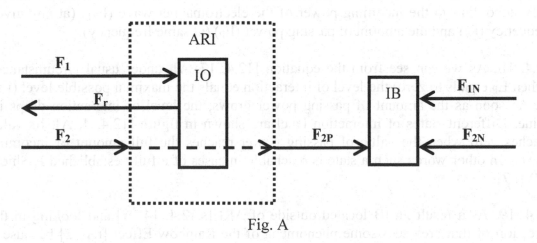

Fig. A

12.4. 15. In figure [12.4. 14. A], F_1 is the electromagnetic wave from the radar (with frequency F_1). F_r is the reflected signal from the target with frequency F_r (in general case $F_1 \neq F_r$ because of target motion). Under usual circumstances the balance of electromagnet power (at the point of the target location) can be described as follows.

$$E_{F1} = E_{Fr} + E_A \tag{a}$$

Where the equation E_{F1} is power of incoming radar signal, E_{FR} is power of reflecting signal, and E_A is absorption power that is collected by the target. As a result the reflecting signal power is less than the power of an incoming radar signal ($E_{FR} < E_{F1}$). But the equation shows the requirements of the law of conservation because the full amount of energy before and after the event (reflection of electromagnet signal) remains constant.

12.4. 16. If a target exists at ARI, only part of the incoming radar signal power interacts with the target, reducing the incoming power of a radar signal. Hence, equation [12.4. 15. A] can be rewritten the following way.

$$E_{F1} = E_{Fr} + E_A + E_P \tag{a}$$

In the equation the additional item, E_P, means passing power. That amount of power passes through ARI without interaction between the electromagnet wave with a given frequency and anything located there (including an IO). Equation [12.4. 16. A] meets the requirements of the law of conservation despite incoming energy rearranging itself.

12.4. 17. That leads us to the estimation of interaction level inside ARI. The following equation shows that mathematically.

$$I(P, F) = \frac{E_{F1} - E_P}{E_{F1}} \tag{a}$$

The equation shows the level of interaction (I) at any given point (P) (mentioned as I(P)) according to the incoming power of the electromagnet wave (E_{F1}) (at any given frequency (F)) and the amount of passing power (E_P) (at same frequency).

12.4. 18. As we can see from the equation [12.4. 17. A], under usual circumstances (when E_P equals to zero) the level of interaction equals the maximal possible level (i.e., 1). As soon as the amount of passing power grows, the level of interaction drops its value. Different states of interaction level are shown in figure [12.4. 4. A]. Its value reaches zero when the value of passing power reaches the full amount of incoming power. In other words such a state is reachable in cases of a fully established E-Shield [r.9.1. 7].

12.4. 19. As a result an IB located outside of ARI [s.12.4. 14. A] and looking in the direction of that area sees some phenomena of the Rainbow Effect [r.12.2] because it can see frequency F_2 unchanged by its interaction with anything located at the region and frequency of F_1 that changed its power (at least) by such interaction. Same frequencies (F_{1N} and F_{2N}) reaching the IB from different directions show no Rainbow Effect.

12.4. 20. The same phenomena help radar signals with different frequencies (F_2) to pass through ARI without losing any power (reflecting, etc.; because I(P,F_2)=0). As a result the radar sees nothing at the area despite possible detection of its signal at the point of an IB's location.

12.5 Operation of Magnetic Compasses

12.5. 1. Another example of an onboard navigation device used widely on aircrafts and sea vehicles is the magnetic compass. "Compass in navigation or surveying, the primary device for direction-finding on the surface of the Earth. Compasses may operate on magnetic or gyroscopic principles or by determining the direction of the Sun or a star."[1]

12.5. 2. "The reason magnetic compasses work as they do is that the Earth itself acts as an enormous bar magnet with a north-south field that causes freely moving magnets to take on the same orientation."[2]

12.5. 3. Generally the operation of magnetic compass is caused by the interaction between Earth's magnetic field and the magnetic field of compass needle. Any deviation of the compass needle causes some force that moves the needle back. As a result the magnetic compass keeps the same orientation of the needle as long as no disturbance exists in the magnet field surrounding the device.

12.5. 4. In the case of a presence of magnetic disturbance around a magnetic compass, the device changes its reading, moving its needle. A changed direction of the needle causes navigational mistakes for anybody who uses the disturbed compass. The most famous cause for the disturbance of a magnetic compass is the presence of an magnetic anomaly.

12.5. 5. "A marine magnetic anomaly is a variation in strength of the Earth's magnetic field caused by magnetism in rocks of the ocean floor. Marine magnetic anomalies typically represent 1 percent of the total geomagnetic field strength. They can be stronger ('positive') or weaker ('negative') than the average total field. Also, the magnetic anomalies occur in long bands that run parallel to spreading centres for hundreds of kilometres and may reach up to a few tens of kilometres in width.

---

[1] "Compass." Encyclopaedia Britannica. Encyclopaedia Britannica 2008 Deluxe Edition. Chicago: Encyclopaedia Britannica, 2008.
[2] "Compass." Encyclopaedia Britannica. Encyclopaedia Britannica 2008 Deluxe Edition. Chicago: Encyclopaedia Britannica, 2008.

"Marine magnetic anomalies were first discovered off the coast of the western United States in the late 1950s and completely baffled scientists. The anomalies were charted from southern California to northern Washington and out several hundred kilometres."[1]

12.5. 6. Hence, any disturbance of the magnetic field around a compass leads to changes in its reading. Moreover, such a disturbance can be made artificially if someone put a magnet next to the compass. Any bar magnet changes the compass readings in that way.

That is the typical response of any magnetic compass made to interact with a surrounding magnetic field (a healthy compass).

12.5. 7. If the needle of a compass slowly loses its magnetization, it causes no result for compass reading until level of the Earth's magnetic force (MF) drops below the level of other natural forces that disturb the magnet needle. In such cases the orientation of the compass caused by the interaction between the needle and other forces (disturbing forces, (DF), magnetic and nonmagnetic one). Figure [12.5. 7.A] shows that graphically.

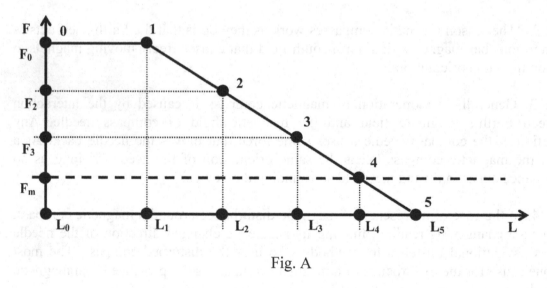

Fig. A

Figure [12.5. 7.A] looks like [12.4. 4.A] with the only difference being that axis P is replaced by axis F (force of interaction between compass magnet needle and surrounding magnetic field). F_0 is force of interaction between a compass needle with full magnetization and the Earth's magnetic force (MF).

Point 1 represents the state of needle magnetization when it begins to lose magnetism. At that point MF equals its usual value (F_0). When the magnetization loses further, it reaches points 2, 3, and 4 in order. Each time the MF reduces to value F_2, F_3, and F_m accordingly.

---

[1] "Ocean." Encyclopaedia Britannica. <u>Encyclopaedia Britannica 2008 Deluxe Edition.</u> Chicago: Encyclopaedia Britannica, 2008.

At point 4 the MF reaches its minimal acceptable limit for the compass's proper operation because DF becomes equal (in magnitude) to MF. As a result the direction of the compass needle at that point depends equally on MF and DF.

Going further between points 4 and 5, MF becomes less than DF, and the reading of the compass needle becomes out of relation with Earth's magnetic field lines. In that state the direction of the compass needle becomes unusable. That happens because any little DF causes casual deviations of needle orientation. At point 5 the needle magnetization becomes equal to zero and the compass stops operating completely.

12.5. 8. The same result can be reached by means of reducing the strength of a surrounding magnetic field. Reducing the magnetic field produces a lesser force on the compass needle, and the less effective the needle becomes. As a result figure [12.4. 4.A] can be applied to such cases without any change. Point 4 represents the state where the force produced by MF becomes almost equal to DF.

12.5. 9. The difference between both ways leads to the same result anyway; reducing the strength of the needle's magnet field or the Earth's magnet field causes the compass to stop working.

12.5. 10. In cases of a magnetic compass in ARI, we have same effect. ARI reduces the interaction between the magnetic field of a compass needle and Earth's magnet field. As a result figure [12.4. 4.A] becomes applicable for such cases as well.

12.5. 11. Hence, ARI reduces the interaction consequently from point 1 to point 5. At point 4 the interaction between Earth's magnetic field and the needle's magnetic field becomes almost equal. Below that point the orientation of the magnetic needle becomes unusable. At point 5 the magnetic compass stops operating completely because the level of interaction becomes equal to zero.

12.5. 12. As we can see all three ways cause a malfunction of the magnetic compass. Any observer who watches the magnetic compass in that state reports a malfunction of the compass because the magnetic needle changes its orientation casually.

12.5. 13. Going from a static magnet field to electromagnet waves, we can see the following relationship between them. A static magnet field can be understood (or imagined) as an electromagnet wave with a frequency equal to zero and a wavelength equal to infinity ($f = c / \lambda$. [s.12.3. 5]).

12.5. 14. Therefore, figure [12.2. 13.A] showing a broken line for LSRE [s.12.2. 14] can be applied for a static magnetic field because that state of field equals point F_0 (zero frequency). In such cases LSRE causes a visible malfunction of magnetic compasses but permits radio contact on frequency F_4 (for example, and any higher frequency) with a vehicle that has the broken magnetic compass on board.

12.5. 15. The difference between a real malfunction of a magnetic compass and the location of the vehicle in ARI is this. If a vehicle goes out of ARI, the magnetic compass suddenly starts to work. The diagram shows that graphically.

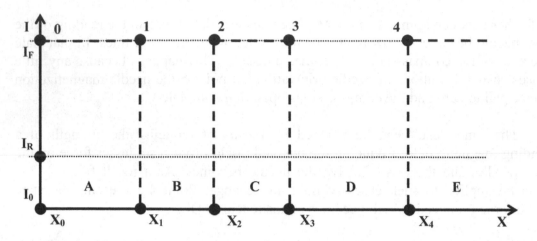

Fig. A

Figure [A] shows the level of interaction between Earth's magnetic field and the magnet needle of a compass. I_F shows the level of full interaction, I_R shows the level of reduced interaction below which the compass has a casual reading, and I_0 shows a zero level of interaction. X_N shows different locations of the moving vehicle.

In areas A, C, and E (X_0-X_1; X_2-X_3; X_4-infinity), a magnetic compass works correctly because the level of interaction equals the maximal possible value (I_F). In areas B and D (X_1-X_2; X_3-X_4), a magnetic compass has casual readings and looks broken.

12.5. 16. As a result a moving vehicle with an onboard magnetic compass can use that device correctly only in areas A, C, and E. Areas B and D cannot be navigated by a magnetic compass because of its casual reading (malfunction).

12.5. 17. If a moving vehicle relocates from area B to C, all onboard magnetic compasses suddenly start to properly work.

12.6 Operation of Gyrocompasses

12.6. 1. Unlike magnetic compasses, gyrocompass need not interact with the magnetic field of the Earth. "The direction a gyrocompass points is independent of the magnetic field of the Earth and depends upon the properties of the gyroscope and upon the rotation of the Earth. The axis of a free gyroscope will describe a circle around the pole of the heavens. To convert it into a gyrocompass, a control must be introduced that, when the axis tilts, will operate to precess (turn) it toward the meridian. The case of the gyroscope is made pendulous, or a liquid is arranged to flow from side to side. Either will convert the path traced by the axis into an ellipse. By delaying the flow of the

100

liquid or by making eccentric the point of action of the control, a damping factor is introduced that converts the ellipse into a spiral so that the gyrocompass eventually settles pointing true north."[1]

12.6. 2. Hence, the main difference between a magnetic compass and a gyrocompass is that the magnetic compass interacts with the magnetic field of the Earth. Any disturbance of that field changes the reading of such a device. A gyrocompass needs only itself to operate correctly. Generally the orientation of a gyrocompass is caused by the interaction of the inner parts of the device (gyroscope) with themselves.

12.6. 3. "Mechanical gyroscopes are based on a principle discovered in the 19th century by Jean-Bernard-Leon Foucault, a French physicist who gave the name gyroscope to a wheel, or rotor, mounted in gimbal rings … The angular momentum of the spinning rotor caused it to maintain its attitude even when the gimbal assembly was tilted. During the 1850s Foucault conducted an experiment using such a rotor and demonstrated that the spinning wheel maintained its original orientation in space regardless of the Earth's rotation. This ability suggested a number of applications for the gyroscope as a direction indicator, and in 1908 the first workable gyrocompass was developed by the German inventor H. Anschutz-Kaempfe for use in a submersible. In 1909 the American inventor Elmer A. Sperry built the first automatic pilot using a gyroscope to maintain an aircraft on course. The first automatic pilot for ships was installed in a Danish passenger ship by a German company in 1916, and in that same year a gyroscope was used in the design of the first artificial horizon for aircraft.

12.6. 4. "Gyroscopes have been used for automatic steering and to correct turn and pitch motion in cruise and ballistic missiles since the German V-1 missile and V-2 missile of World War II. Also during that war, the ability of gyroscopes to define direction with a great degree of accuracy, used in conjunction with sophisticated control mechanisms, led to the development of stabilized gunsights, bombsights, and platforms to carry guns and radar antennas aboard ships. The inertial guidance systems used by orbital spacecraft require a small platform that is stabilized to an extraordinary degree of precision; this is still done by traditional gyroscopes, though mechanical systems are being replaced by optical gyroscopes … Larger and heavier devices called momentum wheels (or reaction wheels) also are used in the attitude control systems of some satellites."[2]

12.6. 5. Independence from any disturbance (or interaction with anything) causes the ability of a gyroscope (and any device using it to orientate itself) to work precisely regardless of the presence or absence of any type of disturbing force of any kind.

12.6. 6. Such an ability leads the gyroscope to work regardless of its location at any point of CE-space or ARI (Gap-Hood) because reducing or even rising interaction

---

[1] "Navigation." Encyclopaedia Britannica. Encyclopaedia Britannica 2008 Deluxe Edition. Chicago: Encyclopaedia Britannica, 2008.
[2] "Gyroscope." Encyclopaedia Britannica. Encyclopaedia Britannica 2008 Deluxe Edition. Chicago: Encyclopaedia Britannica, 2008.

causes no changes on the gyroscope's orientation. As a result, unlike magnetic compass, any device using a gyroscope (gyrocompass, etc.) continues to work properly in CE-space and in ARI as well

12.6. 7. Basically that happens because the gyroscope uses no interaction between itself and the surrounding world. That leads to an absence of any changes in gyroscope operation based on any location (CE-Space, HE-Space, or Cross-Trajectory). In other words, a gyroscope put in ARI or even behind an E-Shield (on HE-Space) does not try to establish cross-shield interaction as the other devices mentioned above. That is a unique device that needs only itself to operate correctly. That leads us to following very important conclusion.

12.6. 8. *Any device that uses no interaction between itself and the physical units of its surrounding world retains correct operation regardless of the presence (or absence) of ARI.* (Gyroscope is a good example of such device.)

12.6. 9. Is there any other example of devices using the same principles of operation? There is one more type of device at least that uses only inner its physical processes for its operation: the clock.

12.7 Operation of Watches and Clocks

12.7. 1. **Watch.** "Watch is a portable timepiece that has a movement driven either by spring or by electricity and that is designed to be worn or carried in the pocket.

12.7. 2. "The first watches appeared shortly after 1500, early examples being made by Peter Henlein, a locksmith in Nurnberg, Ger. The escapement used in the early watches was the same as that used in the early clocks, the verge. Early watches were made notably in Germany and at Blois in France, among other countries, and were generally carried in the hand or worn on a chain around the neck. They usually had only one hand for the hours."[1]

12.7. 3. **Clock.** "Clock is a mechanical or electrical device other than a watch for displaying time. A clock is a machine in which a device that performs regular movements in equal intervals of time is linked to a counting mechanism that records the number of movements. All clocks, of whatever form, are made on this principle."[2]

12.7. 4. "In 1929 the quartz crystal was first applied to timekeeping; this invention was probably the single greatest contribution to precision time measurement. Quartz crystals oscillating at frequencies of 100,000 hertz can be compared and frequency differences determined to an accuracy of one part in 10^{10}.

---

[1] "Watch."Encyclopaedia Britannica. Encyclopaedia Britannica 2008 Deluxe Edition. Chicago: Encyclopaedia Britannica, 2008.
[2] "Clock." Encyclopaedia Britannica. Encyclopaedia Britannica 2008 Deluxe Edition. Chicago: Encyclopaedia Britannica, 2008.

12.7. 5. "The timekeeping element of a quartz clock consists of a ring of quartz about 2.5 inches (63.5 mm) in diameter, suspended by threads and enclosed in a heat-insulated chamber. Electrodes are attached to the surfaces of the ring and connected to an electrical circuit in such a manner as to sustain oscillations. Since the frequency of vibration, 100,000-hertz, is too high for convenient time measurement, it is reduced by a process known as frequency division or demultiplication and applied to a synchronous motor connected to a clock dial through mechanical gearing. If a 100,000 hertz frequency, for example, is subjected to a combined electrical and mechanical gearing reduction of 6,000,000 to 1, then the second hand of the synchronous clock will make exactly one rotation in 60 seconds. The vibrations are so regular that the maximum error of an observatory quartz-crystal clock is only a few ten-thousandths of a second per day, equivalent to an error of one second every 10 years."[1]

12.7. 6. **Atomic clock.** "Atomic clock is type of clock that uses certain resonance frequencies of atoms (usually cesium or rubidium) to keep time with extreme accuracy. The electronic components of atomic clocks are regulated by the frequency of the microwave electromagnetic radiation. Only when this radiation is maintained at a highly specific frequency will it induce the quantum transition (energy change) of the cesium or rubidium atoms. In an atomic clock these quantum transitions are observed and maintained in a feedback loop that trims the frequency of the electromagnetic radiation; like the recurrent events in other types of clocks, these waves are then counted."[2]

12.7. 7. **Nuclear clock.** "Nuclear clock is frequency standard (not useful for ordinary timekeeping) based on the extremely sharp frequency of the gamma emission (electromagnetic radiation arising from radioactive decay) and absorption in certain atomic nuclei, such as iron-57, that exhibit the Mossbauer effect. The aggregate of atoms that emit the gamma radiation of precise frequency may be called the emitter clock; the group of atoms that absorb this radiation is the absorber clock. The two clocks remain tuned, or synchronous, only as long as the intrinsic frequency of the individual pulses of gamma radiation (photons) emitted remains the same as that which can be absorbed. A slight motion of the emitter clock relative to the absorber clock produces enough frequency shift to destroy resonance or detune the pair, so absorption cannot occur. This allows for a thorough study at very low velocities of the Doppler effect (the change in the observed frequency of a vibration because of relative motion between the observer and the source of the vibration). Gamma photons from an emitter placed several stories above an absorber show a slight increase in energy, the gravitational shift toward shorter wavelength and higher frequency predicted by general

[1] "Clock." Encyclopaedia Britannica. Encyclopaedia Britannica 2008 Deluxe Edition. Chicago: Encyclopaedia Britannica, 2008.
[2] "Atomic clock." Encyclopaedia Britannica. Encyclopaedia Britannica 2008 Deluxe Edition. Chicago: Encyclopaedia Britannica, 2008.

relativity theory. Some pairs of these nuclear clocks can detect energy changes of one part in 10^{14}, being about 1,000 times more sensitive than the best atomic clock."[1]

12.7. 8. As we can see, each type of time-keeping device (here and later referred to as TKD) has the same basic principle of operation: the presence of a core process where recurrent events occur. As a result such a type of device does not need to have any interaction with the physical units of the surrounding world (CE-Space). That leads us to the following important conclusion.

12.7. 9. Reading of any TKD depends only on the recurrent physical process that occurs inside the device and its initial indication.

12.7. 10. If a device needs any type of interaction with physical units of the surrounding world (CE-Space) for it to correctly work, it cannot be recognized as a TKD. For example, a clock with a pendulum needs a strong interaction with the gravity field of the Earth to correct work. In case of the absence of a gravity force, such a device is unable to operate.

12.7. 11. Usually TKD is used to indicate current time. And what is time? "Time is a measured or measurable period, a continuum that lacks spatial dimensions. Time is of philosophical interest and is also the subject of mathematical and scientific investigation."[2]

12.7. 12. Here and later I use the first type of the mentioned [12.7. 11] explanations of time. I mean: "Time is a measured or measurable period". Time is more accurate accordingly to purpose of this work.

12.7. 13. According to [12.7. 9], there is no TKD that measures time itself. Any TKD can be used only to estimate a period of time. To keep the reading of TKD suitable, usually they are used with same initial indication. That ensures a similar reading of different TKD for considerable periods of time.

12.7. 14. **Example.** There are two TKD, the chronometer and the watch. At unknown points of time, the watch stopped its operation because of exhausting the power of the mainspring. After some time someone noticed the malfunction of the watch and would like to bring it back to correct operation. The person needs to do following steps:

    a. Restore power of the mainspring
    b. Set the watch to the time of the chronometer

---

[1] "Nuclear clock." Encyclopaedia Britannica. Encyclopaedia Britannica 2008 Deluxe Edition. Chicago: Encyclopaedia Britannica, 2008.
[2] "Time." Encyclopaedia Britannica. Encyclopaedia Britannica 2008 Deluxe Edition. Chicago: Encyclopaedia Britannica, 2008.

12.7. 15. Those steps [12.7. 14.a–b] are enough to bring the watch back to its correct operation. Step A must be done to start the inner, recurrent physical process of the watch (which consumes the power of the mainspring).

12.7. 16. Step B [s.12.7. 14.b] must be done because no TKD has direct information about the current moment of time. Generally any reading difference (precision of time keeping) between two or more TKD is caused only by the difference of recurrent physical process's accuracy and precision of its counting mechanism (which transforms the recurrent physical process to its TKD reading).

12.7. 17. The main features of TKD (mentioned above) let that device operate regardless of its location (CE-Space, HE-Space, or Cross-Trajectory) because it uses only an inner physical processes (that acts equally at CE-Space, HE-Space, or Cross-Trajectory [r.8.5. 1. a-e]) to their operation.

12.8 Transposition Time and Altered Time

12.8. 1. As shown previously [12.7. 9], all TKD are able to count only oscillations of some physical process, not time itself, and they can be used to find time intervals [s.12.7. 17] regardless of trajectory type (RWT or Z-Trajectory). Such aspects can be used to estimate some characteristics of Z-Trajectory.

12.8. 2. Figure [11.1. 1.A] shows some locations of a body moving in an elliptical trajectory. In the case of location A of the moving body and zero distance relocation [s. 11.1. 19.a] ($A_L(t_1)-A_L(t_2)$ relocation, for example), time of such a relocation can be calculated by [10.2. 12.b] and be equal to number of full time revolutions of the moving body around the central one.

12.8. 3. This raises a very important question. Does time of relocation, looking from the point of view of IB [s.2.1. 3.a] (an IS that associates himself (itself) with moving body), and time of relocation looking from the point of view of IO (that uses Z-Trajectory) seem equal?

12.8. 4. In cases of equality of those periods of time, previously synchronized TKDs of IO and IB have the same readings before Z-Trajectory as well as after it. In such cases the TKD of IO (at ZT) has same readings as at RWT.

12.8. 5. Otherwise the readings of IO's TKD (independent observer's time keeping device) become different from the readings of IB's TKD (independent bystander's time keeping device). Moreover, the less distance passed of the moving object (system) by Z-Trajectory, the less time that is used (by the system) to go through Z- trajectory. That happens because of [11.2. 12–13; 11.2. 16]. (Z-Trajectory itself is unable to change the energy of a moving system (object).)[1]

---

[1] This is a law of conservation restriction.

105

12.8. 6. As mentioned in [11.2. 22], ZT-Time always has a minimal calculable value that depends on the length and velocity of the moving system (equation [11.2. 21.B]). Therefore, the difference between readings of IB's TKD and IO's TKD can be found by following equation.

$$T_D = \Delta T_{IB} - \Delta T_Z \qquad \text{(a)}$$

Where T_D is time difference between IB's TKD and IO's TKD (according to their indications), ΔT_{IB} is difference between reading of IB's TKD before and after the event (using Z-Trajectory by the physical system of IO), and ΔT_Z is difference between reading of IO's TKD before and after the event (using Z-Trajectory by the physical system; here and further known as the Z-Event). Before Z-Event the TKD of IO and IB were precisely synchronized and have the same readings.

12.8. 7. In the case of a minimal length of Z-Trajectory, the time difference between IB's TKD and IO's TKD can be calculated by the following equation (transformation of [12.8. 6.A])

$$T_D = \Delta T_{IB} - \frac{2 \cdot L}{V} \qquad \text{(a)}$$

T_D and ΔT_{IB} have the same meaning as they did in equation [12.8. 6.A]; L and V have the same meaning as they did in equation [11.2. 21.b].

12.8. 8. **Example.** If the full trajectory of a moving system is only RWT, ZT-Time equals zero because of the absence of ZT itself. In such cases equation [12.8. 7.a] transforms to following one: $T_D = \Delta T_{IB}$ with time difference between IB's TKD and IO's TKD equal zero. That is the usual state for any moving system. Also that is the main cause for IO's TKD that displays time equal to IB's TKD readings at the first point of RWT and the last one (IO uses RWT).

12.8. 9. In cases when RWT divides for two parts by Z-Trajectory with minimal possible ZT-Time, time delay appears at the end point of RWT. Generally such a difference appears at the moment of time when the moving system begins going through In-Gap. After In-Gap the system exists in CE-Space with a different reading of the TKD.

12.8. 10. In the case of $A_L(t_1)$-$A_L(t_2)$ relocation, for example [s. 11.1. 19.a], $T_D = \Delta T_{IB} = n \cdot T$ (T is a period of revolution). But ZT-Time cannot be equal to zero, and the correct equation becomes the following one: $T_D = \Delta T_{IB} = n \cdot T - (2 \cdot L) / V$.

If ΔT_{IB} is any given period of time according to IB TKD and T_D is any given period of time according to IO TKD, we can calculate ZT-Time by the following equations.

$$T_D = \Delta T_{IB} - \frac{2 \cdot L}{V} \qquad \text{(a)}$$

$$\frac{2 \cdot L}{V} = T_Z \qquad \text{(b)}$$

$$T_D = \Delta T_{IB} - T_Z \qquad \text{(c)}$$

$$T_Z = \Delta T_{IB} - T_D \qquad \text{(d)}$$

12.8. 11. In the case of [1.1], the calculated ZT-Time for aircraft type Boeing 727 traveling at its usual speed has a value calculable by the same equation: $T_Z = (2 \cdot L) / V$. Using the characteristics of that aircraft accordingly, the following table shows we can calculate T_Z for such a type of aircraft.

Table A

Aircraft type	Boeing 727-100
Typical Cruise Speed at cruise altitude (35,000 feet)	972 km/h (0.81 Mach)
Overall length	40.6 m

12.8. 12. Recalculating the speed from [km/h] to [m/s] gives us the following value.

$$972 \text{ [km/h]} = 972*1000 \text{ [m/h]} = 972,000 \text{ [m/h]} = (972,000 / 3,600) \text{ [m/s]} = 270 \text{ [m/s]} \qquad \text{(a)}$$

12.8. 13. Using known values we can calculate T_Z for aircraft type Boeing 727 in the following way.

$$T_Z = (2 \cdot L)/V = (2 \cdot 40.6) / 270 = 0.3 \text{ [s]} \qquad \text{(a)}$$

An exact calculation gives us a value of less than half a second. Even if that aircraft had significantly less speed (half of maximal speed) because of a glide path (landing trajectory of aircraft), T_Z was a bit more than half a second (0.6 sec. at speed 135 [m/s]).

12.8. 14. Hence, the whole process of relocation by Z-Trajectory had displaced it by about half a second (because of the glide path). As mentioned previously [8.5. 1. a–e] there is no difference between processes on CE-Space and HE-Space because of the law of conservation (any change needs additional energy). Therefore, any onboard physical process had displaced only during half a second as well as doubling by a full length the relocation of the aircraft itself.

12.8. 15. Any physical process inside an onboard TKD occurred for same period of time. As a result their readings looked equal after Z-Event as well as before it.

12.8. 16. For other people (located at the airport; IB), the Z-Event had affected accordingly the physical processes of the surrounding world (CE-Space). Because those processes had no relation with HE-Space, they behaved in the usual way. Moreover, any physical process occurred inside any TKD of IB was equal to each other and caused the same readings for their TKD before Z-Event as well as after it (the usual state for any TKD).

12.8. 17. As a result any TKD located in CE-Space had a number of oscillations of its internal physical process that was permitted by CE-Space, and any TKD located in HE-Space had a number of oscillations of its internal physical process that happened by moving the TKD through HE-Space (Z-Trajectory). That variation caused a difference in the readings of the previously synchronized TKDs.

12.8. 18. Usually watches have a precision of not more than one second. As a result they cannot be used for time interval measurement that is less than one second. That feature caused an estimating mistake of the indication of difference between TKDs of the airport and onboard TKDs. According to evidences of the eyewitnesses, the time difference exactly matched the time of absence of the aircraft. That happened because TKD had less precision than was needed to estimate ZT-Time (the difference between 10 minutes and 9 minutes 59 seconds).

12.8. 19. Here and later I refer to the period of time noticed by IB as time of absence of a system (object) that uses Z-Trajectory as Transposition time (T-Time; TT). The indication difference between TKD of IO uses Z-Trajectory (ZT-Time) and IB keeps its location at CE-Space, and I refer to this as Altered time (A-Time; AT). The equation shows [12.8. 6.a] rewritten by those notions.

$$T_A = \Delta T_T - \Delta T_Z \qquad\qquad (a)$$

Where T_A is altered time, T_T is transposition time, and T_Z is ZT-Time.

12.8. 20. The same calculation can be used to find ZT-Time for any system (object, vehicle, etc.) going forward by any given velocity. It's quite possible to calculate ZT-Time using the velocity and length of a vehicle. Such a calculation mentioned below [12.8. 23] is an example of ZT-Time estimation for a large vehicle moving with low speed (tanker).

12.8. 21. "Tanker is a ship designed to carry liquid cargo in bulk. Its cargo is usually a petroleum product, either crude oil being carried from oil fields to refineries or gasoline being carried from refineries to distribution centres. The liquid is piped into the cargo space of the ship and transported without the use of barrels or other containers. Special tankers carry other liquids such as molasses, asphalt, wine, or liquefied natural gas. Tankers differ from general cargo ships in that they normally carry a full load in one direction and return without cargo.

12.8. 22. "Tankers vary in size from small coastal vessels about 60 m (200 feet) long, carrying from 1,500 to 2,000 tons deadweight, up to huge vessels that are more than 500,000 tons deadweight, reach lengths of more than 400 m (1,300 feet), and are the largest ships afloat. Experience with these supertankers has shown that the direct cost of transporting oil goes down as the size of the tanker increases, apparently without limit; an obstacle to building larger tankers is the lack of suitable shore facilities for them."[1]

12.8. 23. **Example**. In the case of a tanker 400 meters in length cruising at low speed equal to 10 km/h (2.78 m/sec), ZT-Time calculation gives the following result.

$$T_Z = (2 \cdot L)/V = (2 \cdot 400) / 2.78 = 287.8 \text{ sec} = 4 \text{ min. } 47.8 \text{ sec} \qquad (a)$$

12.8. 24. In cases when transposition time is equal to 10 minutes (600 sec.), exactly like the previous example, altered time becomes equal to the following [s.12.8. 19.a]:

$$T_A = 600 - 287.8 = 312.2 \text{ sec} = 5 \text{ min. } 12.2 \text{ sec.} \qquad (a)$$

Therefore, if the aircraft from the previous example was replaced by the tanker according to the state described above [12.8. 23], the difference between TKD readings of IB and IO would be equal to 5 minutes, 12.2 seconds (instead of 9 minutes, 59 seconds).

12.8. 25. In other words, the "absence" of the tanker for ten minutes caused a readings difference for 5 minutes, 12.2 seconds, because the tanker took 4 minutes, 47.8 seconds to move in and out of HE-Space. That is the minimal time that must be spent for the process of relocation by Z-Trajectory.

12.8. 26. **Conclusion**. The process of Z-Trajectory used by any system is only possible in cases when ZT-Time equals some determined time that depends on the length and velocity of the system. That time must be more than zero and less that infinity.

12.8. 27. That [s.12.8. 26] is possible only for a system where the relative velocity between Shield-Gap and the system itself exceeds zero. The next topic discusses that subject.

12.9 Spin Effect

12.9. 1. The previous topic discussed physical aspects of ARI and Shield-Gaps according to Gaps that are positioned motionlessly relative to the Earth's surface. As a result ZT-Time was calculated according to a moving system's speed (relative to Earth surface). In the case of moving Shield-Gaps, its velocity must be added to the velocity

---

[1] "Tanker." Encyclopaedia Britannica. <u>Encyclopaedia Britannica 2008 Deluxe Edition</u>. Chicago: Encyclopaedia Britannica, 2008

of the moving system (by law of composition of velocity vectors – two vectors in such a case), and ZT-Time changes dramatically.

12.9. 2. The primary cause for the possibility of presence of moving Shield-Gaps is celestial body rotation. According to figure [11.1. 1.A], a celestial body located at point B has two points (B_1 and B_2) as ZTHP [s.10.2. 17]. Those points can be used as locations of Shield-Gaps [r.5.1. 10. A; r.9.5. 3].

12.9. 3. If the celestial body (planet) has no rotation around its axis, relative velocity between surface of the body and ZTHP equals zero. As a result ZT-Time can be calculated using only the velocity of the moving system because in such a case the relative velocity of system-to-ZTHP and ZTHP-to-body surface look equal.

12.9. 4. If a celestial body rotates around its axis points, B_1 and B_2 poses some value of relative velocity between the points and surface of the body. That is, the space velocity between surface points of the body and an equal gravity strength create a line of the central body (C_m).

12.9. 5. In this case the Earth's maximal velocity appears that way at the equator and can be calculated by the following equation.

$$V_E = \frac{L_E}{T_E}$$
(a)

Where V_E is equatorial velocity of the planet (relative velocity between any points located on equator and axe of the planet), L_E is length of the equator, and T_E is period of equatorial revolution (period of time then an equatorial point proceeds one full revolution).

12.9. 6. The semimajor axis of Earth's length equals 6,378,136±1 m ($a_{equatorial}$ = 6,378,136±1 m).[1] Hence, the length of the equator is equal to $2 \cdot \pi \cdot a_{equatorial}$ = $2 \cdot 3.14 \cdot 6,378,136$ = 40,054,694 m. According to equation [12.9. 5.a] we have following result.

$$V_E = L_E / T_E = 40,054,694 / 24 \cdot 3600 = 463.6 \text{ [m/s]}$$
(a)

That is the exact relative velocity that appears between ZTHP and Earth's surface at zero altitude on zero latitude (the latitude of equator). The time of one full revolution equals 24 hours.

12.9. 7. Therefore, the ZT-Time for a tanker with the characteristics mentioned in [12.8. 23] and located motionlessly at any point with zero latitude (equator) and positioned parallel to the equator equals the following value.

---

[1] "Geoid." Encyclopaedia Britannica. Encyclopaedia Britannica 2008 Deluxe Edition. Chicago: Encyclopaedia Britannica, 2008

$$T_Z = (2 \cdot L) / V = (2 \cdot 400) / 463.6 = 1.73 \text{ [sec]} \qquad \text{(a)}$$

12.9. 8. As we can see, ZT-Time has a very low value that way, despite the significant size of the sea vehicle. That happens because space velocities appear in view in that case.

12.9. 9. As mentioned in [12.1. 9], Gap-Hood must be observed from the outside as "cloud". Hence, Gap-Hood traveling directly from east to west (a motion parallel to equator) can be seen as a "cloud" moving with a high velocity (463.6 [m/s] at zero latitude) in the same direction.

12.9. 10. The higher altitude at which the Gap-Hood takes place, the more velocity it possesses because of the revolving radius rising. Increasing the latitude causes a decreasing velocity of Gap-Hood relative to Earth's surface. At the north and south poles, that velocity becomes zero.

12.9. 11. Here and later I refer to the effect, which appears as relative velocity between any particular point of revolving planet and the ZTHP located motionlessly at a coordinate system bound to the central body, as Spin Effect.

12.10 Z-Process in Gases

12.10. 1. Previous topics [12.8; 12.9] explained the important characteristics of ZT-Time that depend on the velocity and size of a moving system (object). In the case of constant velocity, ZT-Time is higher with the greater length of the moving system (and vice versa).

12.10. 2. Because ZT-Time does not depend on inner physical properties of a moving system, any system has the same rule for ZT-Time calculation despite its size. Here appears one more important question. What is the minimal possible size for a system using Z-Trajectory?

12.10. 3. Under usual circumstances the atmosphere surrounds the Earth. "The atmosphere that surrounds the Earth consists of a mixture of gases, primarily nitrogen and oxygen. This envelope, commonly called the air, also contains numerous less abundant gases, water vapour, and minute solid and liquid particles in suspension."[1]

12.10. 4. "The following is a summary of the above estimates of molecular quantities in a gas, with a little spread in the numbers to allow for molecules both smaller and larger than the typical ones used here – which are H_2O, NH_3, and the nitrogen (N_2) plus oxygen (O_2) mixture that is air – and to allow for the fact that some of these quantities depend on temperature and pressure. It is important to note that these estimates and

---

[1] "Atmosphere." Encyclopaedia Britannica. Encyclopaedia Britannica 2008 Deluxe Edition. Chicago: Encyclopaedia Britannica, 2008.

calculations are rather simplified, although fundamentally correct, and that there may well be missing factors such as $3\pi / 8$ or $\sqrt{2}$. The numerical estimates for gases at ordinary pressure and temperature are:

Molecular diameter	10^{-8} to 10^{-7} cm
Molecular number density	10^{19} molecules/cm^3
Average molecular speed	10^4 to 10^5 cm/s
Average distance between molecules	10^{-7} to 10^{-6} cm
Collision rate per molecule	10^9 to 10^{10} collisions/s
Average time between collisions	10^{-10} to 10^{-9} s
Average distance travel	10^{-5} to 10^{-4} cm

12.10. 5. "The general impression of gas molecules given by these numbers is that they are exceedingly small, that there are enormous numbers of them in even one cubic centimetre, that they are moving very fast, and that they collide many times in one second. Two other facts are especially important. The first is that the lengths involved, especially the mean free path, are minute compared with ordinary lengths, even with the diameter of a capillary tube. This means that gas behaviour and properties are dominated by collisions between molecules and that collisions with walls play only a secondary (though important) role. The second is that the mean free path is much larger than the molecular diameter. Thus, collisions between pairs of molecules are of paramount importance in determining ordinary gas behaviour, while collisions that involve three or more molecules at the same time can basically be ignored."[1]

12.10. 6. It's quite possible to find ZT-Time for a single air molecular, according to aspects of Z-Trajectory, using the physical values mentioned in the previous paragraph. Significant values are the diameter and speed of molecules.

$$T_Z = (2 \cdot L) / V = (2 \cdot 10^{-7}) / 10^4 = 2 \cdot 10^{-11} \text{ [sec]} \qquad (a)$$

Equation A shows that if a molecule has a diameter of 10^{-7} [cm] and a speed of 10^4 [cm/s], then ZT-Time equals $2 \cdot 10^{-11}$ sec. That is less than the average time between collisions (10^{-10} to 10^{-9} s). Comparison between those two periods of time ($2 \cdot 10^{-11}$ sec. and 10^{-10} sec.) leads us to the following conclusion.

12.10. 7. *Relocation of a molecule by Z-Trajectory looks like its motion between two consequent collisions under usual circumstances.*

12.10. 8. That happens because the molecule moves as an independent piece of matter **without** *interaction* with other molecules between two consequent collisions. As a result the presence or absence of Z-Trajectory between two parts of RWT (molecular trajectory between two consequent collisions) produces no changes to the molecular interaction between the molecular itself and other molecules.

---

[1] "Gas." Encyclopaedia Britannica. Encyclopaedia Britannica 2008 Deluxe Edition. Chicago: Encyclopaedia Britannica, 2008

12.10. 9. Therefore, the presence of Shield-Gaps (Out-Gap as well as In-Gap) changes nothing in the behavior of gases surrounding Shield-Gaps. Hence, any physical law that is applicable to gases in their usual state can be applicable to them in the presence of Shield-Gaps and Z-Trajectory itself (if molecules use Z-Trajectory).

12.11 Siren Effect

12.11. 1. There is one physical process in gases that can be noticed directly by the human senses: hearing their sound. "Sound is a mechanical disturbance from a state of equilibrium that propagates through an elastic material medium. A purely subjective definition of sound is also possible, as that which is perceived by the ear, but it is not particularly illuminating and is unduly restrictive, for it is useful to speak of sounds that cannot be heard by the human ear, such as those that are produced by dog whistles or by sonar equipment."[1]

12.11. 2. "Most solids, liquids, and gases of ordinary experience can serve as media for sound. A mechanical disturbance may be produced in any number of ways but will consist of a sudden increase in pressure at some point. Since the material is elastic, the compression is not permanent; once the disturbing influence is removed, the compressed region will rebound, but in doing so it will compress an adjacent region. The result of this cycle repeating itself is the generation of a compression wave, followed by a rarefaction wave as each region of elastic material rebounds. These waves are longitudinal, i.e., the displacement of a particle of the medium is in the direction of wave motion (note, however, that there is no net transport of material). The waves thus generated travel through the medium at a speed that is a function of the equilibrium pressure and density of the material and, to various extents, of the specific heat (for a gas), the elasticity (for liquids and solids), and the temperature of the medium and of the frequency of the wave. In dry air (at 0 °C [32 °F] and a sea-level pressure of 14.7 pounds per square inch [1013.25 millibars]) the speed of sound is 331.29 metres per second (741.1 miles per hour). (This calculation, made in 1986, corrected an earlier calculation made about 1942.) The speed of sound in seawater is 1,490 metres (4,889 feet) per second and in steel 5,000 metres (16,405 feet) per second.

12.11. 3. "Another parameter of sound, in addition to velocity and frequency, is intensity, which is defined as the average flow of energy per unit time through a unit area of the medium. Intensity is usually measured in watts per square centimetre. The standard usually used for the quietest sound audible to the human ear has an intensity of 10^{-16} watt per square centimetre. Closely related to the intensity of a sound is the sound pressure, or pressure excess over equilibrium caused by a sound wave. Sound pressure is measured in dynes per square centimetre or, in the SI system of units, in pascals. The sound pressure of the human voice at ordinary conversational level, measured directly in front of the mouth, is about 0.1 pascal, which may be compared with the standard

---

[1] "Sound." Encyclopaedia Britannica. <u>Encyclopaedia Britannica 2008 Deluxe Edition</u>. Chicago: Encyclopaedia Britannica, 2008.

pressure of the atmosphere of about 105 pascals. Also related to sound intensity, although not simply, is loudness. Loudness is a subjective phenomenon and can be measured only comparatively, by means of a standard reference sound under specified conditions."[1]

12.11. 4. If there is some value of sound intensity [s.12.11. 3] at the Out-Gap point (for example 1 Watt), and the physical process of sound propagation is untouched by Z-Trajectory [s.12.10. 9], then In-Gap possesses the same value of sound intensity (1 Watt). That exactly matches the law of conservation.

12.11. 5. Hence, the presence of IB around an In-Gap that has a corresponding Out-Gap with an audible sound source leads to an acoustic impression to IB at the point of its location. From the IB's point of view, it the IB hears an audible sound without any visible sound source. Obviously, such circumstances lead humans to be nonplussed.

12.11. 6. Because Z-Trajectory is able to have an In-Gap and Out-Gap with a large time difference between them (transposition time), an IB (human) can hear familiar sounds and unrecognizable sounds as well those that were never felt by him or her before. Here and later I refer to that phenomenon as the Siren Effect.

12.11. 7. "Siren in Greek mythology, a creature half bird and half woman who lured sailors to destruction by the sweetness of her song. According to Homer there were two Sirens on an island in the western sea between Aeaea and the rocks of Scylla. Later the number was usually increased to three, and they were located on the west coast of Italy, near Naples. They were variously said to be the daughters of the sea god Phorcys or of the river god Achelous."[2]

12.12 Z-Process in Liquids

12.12. 1. Unlike gases, the molecules of liquids have a different interplay order and location. That causes the difference between physical properties of gases and liquids. In liquids the molecules are much closer to each other than in gases. That helps them to establish some interaction between neighboring molecules. "The most likely location for a second molecule with respect to a central molecule is slightly more than one molecular diameter away, which reflects the fact that in liquids the molecules are packed almost against one another. The second most likely location is a little more than two molecular diameters away, but beyond the third layer preferred locations damp out, and the chance of finding the centre of a molecule becomes independent of position."[3]

[1] "Sound." Encyclopaedia Britannica. Encyclopaedia Britannica 2008 Deluxe Edition. Chicago: Encyclopaedia Britannica, 2008.
[2] "Siren." Encyclopaedia Britannica. Encyclopaedia Britannica 2008 Deluxe Edition. Chicago: Encyclopaedia Britannica, 2008.
[3] "Liquid." Encyclopaedia Britannica. Encyclopaedia Britannica 2008 Deluxe Edition. Chicago: Encyclopaedia Britannica, 2008.

12.12. 2. Such close location and interaction between molecules in liquids causes the possibility of some specific processes that appear only in liquids. One of them is evaporation. "Vaporization also called Evaporation, conversion of a substance from the liquid or solid phase into the gaseous (vapour) phase. If conditions allow the formation of vapour bubbles within a liquid, the vaporization process is called boiling. Direct conversion from solid to vapour is called sublimation.

12.12. 3. "Heat must be supplied to a solid or liquid to effect vaporization. If the surroundings do not supply enough heat, it may come from the system itself as a reduction in temperature. The atoms or molecules of a liquid or solid are held together by cohesive forces, and these forces must be overcome in separating the atoms or molecules to form the vapour; the heat of vaporization is a direct measure of these cohesive forces.

12.12. 4. "Condensation of a vapour to form a liquid or a solid is the reverse of vaporization, and in the process heat must be transferred from the condensing vapour to the surroundings. The amount of this heat is characteristic of the substance, and it is numerically the same as the heat of vaporization."[1]

12.12. 5. Also there is one process that occurs equally in gases and liquids. That is the process of pressure appearance. "Pressure in the physical sciences, is the perpendicular force per unit area, or the stress at a point within a confined fluid. The pressure exerted on a floor by a 42-pound box the bottom of which has an area of 84 square inches is equal to the force divided by the area over which it is exerted; i.e., it is one-half pound per square inch. The weight of the Earth's atmosphere pushing down on each unit area of the Earth's surface constitutes atmospheric pressure, which at sea level is about 15 pounds per square inch. In SI units, pressure is measured in pascals; one pascal equals one newton per square metre. Atmospheric pressure is close to 100,000 pascals.

12.12. 6. "The pressure exerted by a confined gas results from the average effect of the forces produced on the container walls by the rapid and continual bombardment of the huge number of gas molecules. Absolute pressure of a gas or liquid is the total pressure it exerts, including the effect of atmospheric pressure. An absolute pressure of zero corresponds to empty space or a complete vacuum."[2]

12.12. 7. In case of the presence of an Out-Gap in liquids, no liquid is able to use a corresponding Shield-Gap (In-Gap, and Z-Trajectory itself) because "The atoms or molecules of a liquid or solid are held together by cohesive forces" [s.12.12. 3]. In other words, it is impossible for the Z-Process to separate any molecules from any liquid directly. That happens because the separation of a molecule from a liquid consumes some energy (caused by the presence of cohesive forces). As a result the law of

[1] "Vaporization." Encyclopaedia Britannica. Encyclopaedia Britannica 2008 Deluxe Edition. Chicago: Encyclopaedia Britannica, 2008.
[2] "Pressure." Encyclopaedia Britannica. Encyclopaedia Britannica 2008 Deluxe Edition. Chicago: Encyclopaedia Britannica, 2008.

conservation denies such a process (as well as any other process that consumes any type of energy from nowhere).

12.12. 8. Therefore, liquid itself is unable to reach the Out-Gap because there is no last point of flowing liquid [s.11.2. 9]. Each molecule that reaches Out-Gap keeps cohesive forces that bind it to other molecules. That state appears for each molecule that reaches the same Out-Gap. Hence, we have the following law for liquids.

12.12. 9. *Z-Process is not applicable to liquids themselves.*

12.12. 10. Nevertheless the molecules of liquids are able to use Z-Trajectory by means of an additional process that helps them to break cohesive forces and leave a liquid completely. Usually that process is called evaporation [s.12.12. 2–12.12. 4].

12.12. 11. In case of evaporating, molecules (or atoms) of liquids change their condition to a gas. Therefore, the activity of those particles in that changed condition has no difference from the activity of gas particles. As remembers reminder, Z-Process in gases was described at [12.10]. The following figure shows Z-Process in liquids.

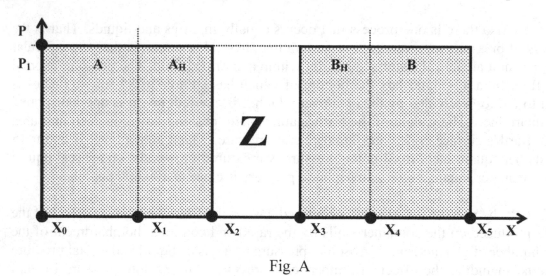

Fig. A

12.12. 12. In figure [12.12. 11.A], X is abscissa; P is axe of pressure. X_N is point of CE-Space; P_1 is level of pressure. There is equal pressure at any location because $P(X_0)=P(X_1)=P(X_2)= \ldots =P(X_5)=P(X_N)$. X_2 and X_3 are points of corresponding Shield-Gaps. The area between X_2 and X_3 is HE-Space (Z-Trajectory itself). It is represented in the figure by the Z letter. Two areas, X_0-X_1 and X_4-X_5, mentioned as A and B, represent parts of CE-Space. Those areas are separated from Shield-Gaps by Gap-Hoods represented in the figure as A_H and B_H areas (X_1-X_2 and X_3-X_4 accordingly).

12.12. 13. The equality of pressure at any point shown in figure [12.12. 11.A] is caused by mechanical interaction between the molecules of liquid. Each molecule exerts pressure on the others as long as the liquid exists. Usually that aspect refers as Pascal's law.

12.12. 14. "Pascal's principle also called Pascal's Law, in fluid (gas or liquid) mechanics, statement that in a fluid at rest in a closed container a pressure change in one part is transmitted without loss to every portion of the fluid and to the walls of the container. The principle was first enunciated by the French scientist Blaise Pascal."[1]

12.12. 15. "Pascal also discovered that the pressure at a point in a fluid at rest is the same in all directions; the pressure would be the same on all planes passing through a specific point. This fact is also known as Pascal's principle, or law."[2]

12.12. 16. As mentioned at [12.12. 6], the main cause for pressure appearance is the continual bombardment of molecules – that is, the molecular motion of gases or liquids. Generally that process is caused by the permanent, chaotic movement of molecules of any liquid or gas.

12.12. 17. Because any moving particle has some kinetic energy, it keeps that amount of energy as long as there is no cause for energy change.[3] Therefore, molecules are able to reach any Shield-Gap as well as any other object. That happens because neither Z-Trajectory nor Gap-Hood is able to change the energy of an incoming piece of matter (object, molecule, particle, etc.).[4] As a result the molecules of a liquid or gas are able to reach the Shield-Gap through Gap-Hood by means of their own energy of motion.

12.12. 18. As soon as a molecule of liquid reaches Shield-Gap, it can go through it, as in case of a solid object. But liquid never reaches the corresponding Shield-Gap [b.12.12. 9]. As a result the liquid keeps its initial position regardless of the presence or absence of a Shield-Gap as well as a solid object that never uses Z-Trajectory because of its slow motion [r.11.2. 14].

12.12. 19. As mentioned in [12.12. 10–12.12. 11], molecules of liquid are able to use Z-Trajectory only in a state of vapor. Each molecule that transfers from a liquid state to a vapor state decreases the energy of liquid (reducing the liquid temperature), which is mentioned in [12.12. 3]: "Heat must be supplied to a solid or liquid to effect vaporization. If the surroundings do not supply enough heat, it may come from the system itself as a reduction in temperature." Therefore, any molecular that goes out of the liquid by Z-Trajectory must reduce the liquid's temperature at the point of Shied-Gap location.

12.12. 20. According to figure [12.12. 11.A], any molecule using Z-Trajectory from the X_2 point to the X_3 point reduces the temperature of the liquid at X_2 point (Out-Gap for the molecule).

---

[1] "Pascal's principle." Encyclopaedia Britannica. Encyclopaedia Britannica 2008 Deluxe Edition. Chicago: Encyclopaedia Britannica, 2008.
[2] "Pascal's principle." Encyclopaedia Britannica. Encyclopaedia Britannica 2008 Deluxe Edition. Chicago: Encyclopaedia Britannica, 2008.
[3] Corollary from the law of conservation.
[4] Corollary from the law of conservation.

12.12. 21. In cases of the opposite direction, a molecule goes to X_3 point from B_H area. It uses Z-Trajectory by the same law and reduces the temperature of the liquid at X_3 point (Out-Gap for the molecule).

12.12. 22. At the end point of Z-Trajectory (In-Gap), a molecule reaches liquid again and hits the surrounding molecules. That process is the same as the process of condensation [s.12.12. 4] and leads to the reduction of the amount of energy of a molecular, transferring that energy (heat) to surrounding molecules – that is, increasing the temperature of the liquid at In-Gap point.

12.12. 23. The amount of energy that a molecular possesses from liquid at the Out-Gap point equals exactly the amount of energy that same molecular transfers to surrounding molecules at the In-Gap point. That process exactly matches the law of conservation. The same law leads to an equality between evaporation and condensation energy [r. 12.12. 4]. That happens because both ways (evaporation-condensation and Z-Process) use the same physical process of extracting molecules from liquid and putting them in the liquid again. Both processes break and establish liquid molecules' cohesive forces by changing their own energy and the energy of liquid.

12.12. 24. Hence, the amount of energy that a molecule possess from liquid at Out-Gap point and transfers to In-Gap point (as energy of the incoming molecular) are equal to each other.

12.12. 25. In cases when the same type of liquid surrounds both corresponding Shield-Gap points and has equal pressure at both Shield-Gap points, the whole system, including the liquid, corresponding Shield-Gaps, and Z-Trajectory, keeps dynamic equilibrium. That happens because the average number of molecules that uses Z-Trajectory both ways keeps constant for any given period of time. That leads us to the following conclusion.

12.12. 26. *In cases of equal pressure and the type of liquid at both corresponding Shield-Gaps, the Z-Process is not noticeable for an independent bystander (IB).*

12.12. 27. If the pressure at corresponding Shield-Gaps becomes unequal to each other, the Z-Process changes significantly. The figure shows that case.

118

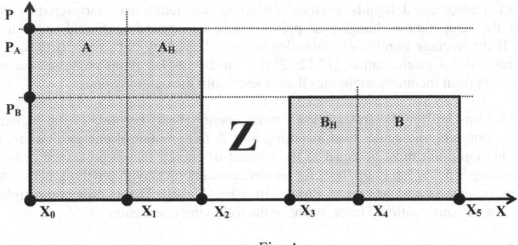

Fig. A

Figure [A] has only one difference from figure [12.12. 11.A], in the value of area B pressure. That is shown in the figure as the different pressure level of P_A and P_B ($P_A > P_B$). Hence, the pressure of area A is greater than the pressure of area B.

12.12. 28. In such a case [12.12. 27], the Shield-Gap located at X_2 point has a greater pressure and a greater number of molecules that attempt to use Z-Trajectory. That looks like increasing the number of molecule bombardment for any surface that is put in the liquid. The more pressure the liquid has, the more molecules bombarding the same area of the surface at the same time.

12.12. 29. In case of the Z-Process increasing pressure at Shield-Gap point, it increases the probability for the existence of molecules that have enough energy to use Z-Trajectory. That leads to an increased number of molecules that use Z-Trajectory at the same time at that particular Shield-Gap.

12.12. 30. Moreover, decreasing the pressure (as at X_3 point [s.12.12. 27.A]) leads to a decreasing probability for the presence of molecules that have enough energy to use Z-Trajectory at the appropriate Shield-Gap.

12.12. 31. *Therefore, in the case of equal types of liquid and different pressures at corresponding Shield-Gaps, the probability of using Z-Trajectory by molecules is higher when there is more pressure at the given Shield-Gap.*

12.12. 32. As a result the average number of molecules that use Z-Trajectory and pass from X_2 to X_3 [s. 12.12. 27.A] is higher than the number of molecules going back (from X_3 to X_2). Such a difference leads to a broken dynamic equilibrium [r.12.12. 24] and produces some "flow" from the Shield-Gap point with higher pressure to the corresponding Shield-Gap point with lower pressure. Such flow appears as a difference in the number of molecules that uses Z-Trajectory frontways and backways. Here and later I refer such a process as the Z-Current.

12.12. 33. Unlike usual liquids' current, Z-Current has additional characteristics. It leads to the temperature changing at points where corresponding Shield-Gaps are located. If the average number of molecules using Z-Trajectory both ways is unequal (in the case of Z-Current, unlike [12.12. 25]), then the In-Gap point possesses more energy (heat) from incoming molecules than it loses with the outgoing one.

12.12. 34. Out-Gap (unlike In-Gap) loses more energy (heat) because of the higher number of outgoing molecules than incoming one. A full amount of energy that loses one Shield-Gap point must be equal to the amount of energy that is possessed in the corresponding Shield-Gap point, because no molecule interacts with anything between the corresponding Shied-Gaps (see Z-Process in gases [12.10]). That is fully compatible with the law of conservation. Hence, we have the following conclusion.

12.12. 35. *Z-current transfers some mass of liquid from Out-Gap to In-Gap and changes the temperature at both Shield-Gap points.*

12.12. 36. In the case of equal temperature of the liquids at two different pools before Z-Current appearance, the pool containing Out-Gap decreases in liquid volume and temperature. The corresponding pool containing In-Gap increases in liquid volume and temperature.

12.12. 37. In the case of different temperature, the actions of Shield-Gaps don't change in the area of liquid mass transferring – that is, Out-Gap always has more outgoing molecules than incoming ones compared to the corresponding In-Gap.

12.12. 38. The area of temperature is uncertain in that the difference between thermal states of liquid at corresponding Shield-Gaps is able to create a different type of result. For example, if the average thermal energy of incoming molecules is less than the thermal energy of molecules at In-Gap, they produce a reduction of temperature at the In-Gap point.

12.12. 39. If the average thermal energy of incoming molecules equals the thermal energy of molecules at the In-Gap point, Z-Process changes no liquid temperature at the In-Gap point (as well as in case of dynamic equilibrium [s.12.12. 25], but it produces Z-Current that way).

12.12. 40. If the average thermal energy of outgoing molecules equals the thermal energy of molecules at the Out-Gap point, Z-Process changes no liquid temperature at the Out-Gap point (as well as in case of dynamic equilibrium [s.12.12. 25], but it produces Z-Current that way).

12.12. 41. In general cases the average thermal energy of Z-Current is unequal for incoming and outgoing molecules at corresponding Shield-Gaps. That leads to a temperature changes at both points (corresponding Shield-Gaps) [r.12.12. 35].

12.12. 42. Therefore, in the case of equal liquid and unequal pressure at corresponding Shield-Gaps, Z-Process leads to an increasing volume of liquid in the pool containing In-Gap both ways by increasing the volume of liquid and its temperature. Hence, the additional volume of liquid at the In-Gap pool appears in Z-Process as a sum of the thermal volume increasing and the mechanical volume increasing by an additional volume of liquid transferred to the In-Gap by Z-Process. The following equation shows that mathematically.

$$\Delta V_Z = \Delta V_T + \Delta V_C \qquad \text{(a)}$$

In equation A, ΔV_Z is additional volume that is added to liquid at In-Gap point by Z-Process, ΔV_T is volume that is added to liquid at In-Gap point by thermal disturbing, and ΔV_C is volume that is added to liquid at In-Gap point by Z-Current.

12.12. 43. In the case of an Out-Gap, equation [12.12. 42. a] transfers to the following one.

$$\Delta V_Z = -\Delta V_T - \Delta V_C \qquad \text{(a)}$$

The meaning of symbols in equations [12.12. 42. a] and [12.12. 43.a] are equal. The only difference is this: in the case of Out-Gap, the volume decreasing equals the volume decreasing by thermal disturbing (ΔV_T) and the volume reducing by losing some liquid through Z-Current (ΔV_C). That leads us to the following conclusion.

12.12. 44. In both ways the volume difference would be more than the volume caused by any process done separately.

12.12. 45. An additional phenomenon related to Z-Current is thermal current (convection). That happened because the number of molecules around S-Gap rises with the more time the corresponding S-Gaps exists. Molecules with different temperature gradually move from S-Gap toward Gap-Hood, and vice versa. That process is usually called diffusion [s.12.12. 47-12.12. 49]. The closer the molecules go to edge of Gap-Hood (X_1 and X_4 [s. 12.12. 27.A]), the more interaction with the gravity field they possess (characteristics of Gap-Hood [r.9.3–9.4]). As a result the thermal current (convection [s.12.12. 50–12.12. 51]) appears at both points (around Out-Gap and In-Gap).

12.12. 46. To those who are not quite familiar with the processes of diffusion and convection, I will provide brief definitions.

12.12. 47. "Diffusion is process resulting from random motion of molecules by which there is a net flow of matter from a region of high concentration to a region of low concentration. A familiar example is the perfume of a flower that quickly permeates the still air of a room.

121

12.12. 48. "Heat conduction in fluids involves thermal energy transported, or diffused, from higher to lower temperature. Operation of a nuclear reactor involves the diffusion of neutrons through a medium that causes frequent scattering but only rare absorption of neutrons.

12.12. 49. "The rate of flow of the diffusing substance is found to be proportional to the concentration gradient. If j is the amount of substance passing through a reference surface of unit area per unit time, if the coordinate x is perpendicular to this reference area, if c is the concentration of the substance, and if the constant of proportionality is D, then $j = -D(dc/dx)$; dc/dx is the rate of change of concentration in the direction x, and the minus sign indicates the flow is from higher to lower concentration. D is called the diffusivity and governs the rate of diffusion."[1]

12.12. 50. "Convection is process by which heat is transferred by movement of a heated fluid such as air or water.

12.12. 51. "Natural convection results from the tendency of most fluids to expand when heated – i.e., to become less dense and to rise as a result of the increased buoyancy. Circulation caused by this effect accounts for the uniform heating of water in a kettle or air in a heated room: the heated molecules expand the space they move in through increased speed against one another, rise, and then cool and come closer together again, with increase in density and a resultant sinking."[2]

12.13 Surge Effect

12.13. 1. In cases of usual circumstances for two IB watching two different pools connected by corresponding Shield-Gaps, the Z-Process appears like this: The pool that has the higher level of liquid possesses a higher pressure at S-Gap level. That happens because the hydrostatic pressure depends on the distance between liquid surface and any given point in the liquid.

12.13. 2. "A column of water, so much denser than air, exerts a greater amount of pressure than a column of air. With each 10-metre (32.8-foot) increase in depth, there is an increase in hydrostatic pressure equivalent to one atmosphere. Mean ocean depth is about 3,800 metres and has a pressure of about 380 atmospheres."[3]

12.13. 3. According to figure [12.12. 27.A], different levels of hydrostatic pressure are shown as P_A and P_B. In such cases the S-Gap located in pool A becomes an Out-Gap

[1] "Diffusion." Encyclopaedia Britannica. Encyclopaedia Britannica 2008 Deluxe Edition. Chicago: Encyclopaedia Britannica, 2008.
[2] "Convection." Encyclopaedia Britannica. Encyclopaedia Britannica 2008 Deluxe Edition. Chicago: Encyclopaedia Britannica, 2008.
[3] "Biosphere." Encyclopaedia Britannica. Encyclopaedia Britannica 2008 Deluxe Edition. Chicago: Encyclopaedia Britannica, 2008.

for liquid because of the greater pressure, and the S-Gap located in pool B becomes an In-Gap because of the lesser pressure of liquid at S-Gap level.

12.13. 4. Hence, Z-Current appears as a relocation of liquid from A pool to B pool. Moreover, the temperature of pool A decreases gradually (and the temperature of pool B increases as well) with the higher the volume of liquid that passes between two pools by Z-Current [s.12.12. 41].

12.13. 5. As a result two different IB watching pools A and B see an unexpected changing of water level of both pools from the time of establishing Z-Current to the time when Z-Current drops to zero as well as closing the S-Gaps. The IB of pool A notices decreasing water pool temperature, and the IB of pool B notices increasing water pool temperature.

12.13. 6. Hence, both IB, who both know modern physics, have decided there was an "unexpected changing of water level caused by thermal disturbance". The difference between thermal disturbance and Z-Current is that Z-Current is able to produce much more liquid (water) level changing than any type of thermal disturbance because it involves a transferring of matter (liquid itself).

12.13. 7. Here and later I refer to the process of the water level quickly changing in any given water pool that coincides with same direction of thermal disturbing (increasing or decreasing water volume) as Surge Effect.

12.13. 8. Such an effect is more noticeable with the is less volume that the pool has (in the case of Z-Current constant value).

12.14 Window Effect

12.14. 1. Any physical appearance of Z-Process and the related processes of Gap-Hoods mentioned above in this chapter have the same characteristics according to the timeline from IB's point of view. The figure shows that graphically.

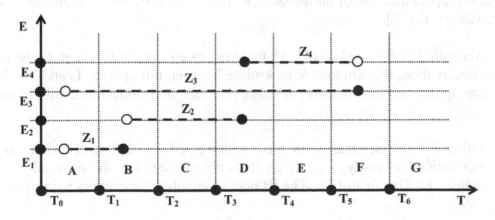

Fig. A

123

Figure [A] shows the diagram with axes of time and events. T is the time axis and E is the event axis. The time axis has some different points of time mentioned as T_0, T_1, T_2, … T_6. Capital letters from A to G show different time intervals according to the nearest time points. For example the time interval D exists between time points T_3 and T_4. If an IB uses any given time interval for observation of any event, that time interval can be called a time window or Observation Window (OW).

12.14. 2. Axis E [s.12.14. 1.A] shows a number of events (E_1, E_2, E_3, E_4,) that happen for some physical systems that use Z-Trajectory. Each system begins Z-Trajectory at the Out-Gap point that is shown as a simple circle, and finishes that trajectory at the In-Gap point that is shown as a filled circle. Each trajectory is marked by the letter Z with a numeric subscript (Z_1, Z_2, Z_3, Z_4).

12.14. 3. As we can see, event E_1 happens for the first physical system and includes its passage trough Z_1 Out-Gap at observation window A, using Z-Trajectory to B observation window B and passage through Z_1 In-Gap. The same sequence of events happens for any physical system that uses Z-Trajectory and for any IO using the same system as well.

12.14. 4. Because IB is able to see only the first and last event of Z-Process (passage through Out-Gap and/or In-Gap), the IB can see only events at the A OW and the B OW. Those events look independent to the IB and appear very "strange". IB sees a strange appearance or disappearance of some objects from nowhere at each OW.

12.14. 5. The first written phenomenon that has broken that unexplainable process was the Boeing [s.1.1]. That event clearly showed that the physical object before and after Z-Process was same one. Even the memories of onboard humans and their personalities were unchanged by the Z-Process.

12.14. 6. That became possible only because there was the same IB (airport staff, radars, etc.) that saw the whole Z-Process from first to last points. Generally that IB, as any other IB who watches Z-Process, can see only the first and last point and the event of Z-Process (appearance and/or disappearance of an object) because Z-Trajectory itself is unobservable [r.6.1. 8].

12.14. 7. According to figure [12.14. 1.A], that process equals the E_1 event. In the case of Boeing observation, the whole OW took place between time points T_0 and T_2. As a result it was quite easy to compare the disappearance and appearance of the same object.[1]

12.14. 8. Other Z-Processes that occur for very long periods of time according to IB TKD are not noticed so easily. Usually an IB is able to see only the physical system passing through Out-Gap or In-Gap. The IB notices the observing object's unexplained

---

[1] And understand the equality of the same object before and after the event.

appearance or disappearance that way. Event E_2 [s.12.14. 1.A] can be used as good example for such an instance of Z-Process.

12.14. 9. E_2 happens when Z-Process appears between Out-Gap located at B OW and In-Gap located at D OW. Hence, any IB using C OW has no idea about the presence of the Z_2-Process that time. As long as the IB keeps observation at C OW, it has no chance to see anything relating to theZ_2-Process [r.6.1. 8].

12.14. 10. The same condition applies to IB accordingly with the Z_3-Process. It is unobservable for the IB at C OW too. Therefore, any number of Z-Processes are unobservable for IB as long as it is not the right point of time when the Z-Trajectory shows its Out-Gap or In-Gap. That leads us to following conclusion.

12.14. 11. *As long as the IB is unable to see (feel, notice, etc.) any Gap of Z-Trajectory and any process related to them, the IB is unable to have any idea about the existence of Z-Process itself.*

12.14. 12. If the IB has an observation between T_3 and T_4 points of time [s.12.14. 1.A] (it relates to D OW), the IB can see Out-Gaps of two Z-Processes (Z_2 and Z_4). If each of those processes transfers an object (or system), IB sees the strange appearance of both objects at the observing area and at the observation time interval (D OW).

12.14. 13. Observation windows C and E have the same characteristics because no Shield-Gap exists at those observation widows.

12.14. 14. Observation window F contains an In-Gap of Z_3-Process and an Out-Gap for Z_4-Process. That process leads the transferring system to corresponding with the In-Gap located at D observation window.

12.14. 15. As well as for any other Z-Processes, Z_4-Process keeps the same rule that is applicable for any Z-Process. In other words, the time difference between TKD readings of IO uses Z_4-trajectroy, and IB TKD can be calculated by the equation [12.8. 19.a]. But that way the value of T_A is hardly usable because from the IB's point of view, IB is unable to estimate future readings of its TKD at the time of Z_4 Out-Gap.

12.14. 16. Nevertheless the IO's TKD has readings at Z_4 In-Gap as its readings at Z_4 Out-Gap, plus Z_T according to [12.8. 10.b]. I mention this very interesting process later on in the work; it helps to explain some unusual phenomena.

12.14. 17. Here and later I refer to the IB's inability to have any evidence according to the Z-Process and the relating processes outside of the appropriate time interval (observation window) as the Window Effect.

12.14. 18. Additionally, I refer to the relocation at a later point of time relative to the IB's TKD as Positive Time Transposition (PTT). The reverse relocation appears as Negative Time Transposition (NTT).

Chapter 13. Biological Appearances

13.1 Z-Trajectory and Living Creatures

13.1. 1. Chapter [12] discusses Z-Phenomena according to a physical system ignoring the life phenomenon. Generally all phenomena mentioned for lifeless systems are applicable to alive systems because any such system (creature) is not more than an example of a physical system. It has mass, size, mechanical properties, and so on. Any of those properties can be found at both types of systems (alive and lifeless).

13.1. 2. For the purposes of this work, I use the notion biological system (BS) to refer to any creature that is recognized by humans as alive (animals, birds, fish, etc.). I use the notion of system (instead of object) because most parts of Earth creatures contains various objects and substances inside their bodies (blood, bones, muscles, specific organs, etc.).

13.1. 3. To simplify further explanations, I refer to the combination of Z-Trajectory, two corresponding gaps (Out-Gap and In-Gap) linked by that trajectory, and two Gap-Hoods (surrounding both gaps) as Z-Gate (ZG).

13.1. 4. In a conservative field BS has the same type of interaction with such a field as with lifeless systems (objects). In the case of a gravity field, any BS feels gravity attraction and interacts with the gravity field.

13.1. 5. In that case, any rule of ZG can be applicable to BS with the same result as compared to any other physical system.

13.1. 6. Such a result can be noticed by humans as an observation of usual (or unusual) creatures at unusual places that are absolutely unreachable to the creatures by any RWT. Hence, such observation can be used as a separation between the possibilities of presence RWT and ZT. In the case of using Z-Gate by any type of BS location, a BS at any unreachable place can be explained only by aspects of ZG.

13.1. 7. As mentioned previously [r.6.1. 8], there is no energy interaction between any systems using Z-Trajectory (here and further called the Passing System, PS) and ZG itself. Hence, if there is no energy coming to or from the Passing System inside Z-Gate, then no biological system has any feeling of the ZG's presence or absence. That happens because any feeling is possible only as a result of an energy interaction between the BS and the surrounding world.

13.1. 8. **Example**. The feeling of heat is possible only when a creature absorbs some heat (thermal energy) from the surrounding world. The feeling of cold is possible only when a creature radiates some heat (thermal energy) to the surrounding world. Acoustic sense is only possible when a creature absorbs some energy of acoustic waves through its sound-sensitive organs (and so on and so forth).

That example can be extended for any sense of any creature. That leads us to the following important conclusions.

13.1. 9. Any biological system must have some energy interaction with the surrounding world to have any feeling of that world.

13.1. 10. No biological system has any feeling of the Z-Gate passage because a Z-Gate produces no energy interaction with any passing system.

13.1. 11. Some feelings for the BS are possible only at the area of Gap-Hood (ARI) if the creature is able to feel electromagnetic waves. Also, the BS is able to feel the Siren Effect at a close location to the Gap-Hood.

13.2 Dragon Effect

13.2. 1. To find possible evidence of a Z-Gate that was used by BS, we need to understand the S-Outline looking from the Earth's surface. In other words, what is the S-Outline according to the Earth's surface?

13.2. 2. To answer that question, we need to remember [7.3. 1.A; 7.3. 4] (Circle D_1-D_2 at figure 7.3. 1.A). In that case CB_a represents the Earth, and CB_b represents the sun. Accordingly the Earth surface D_1-D_2 circle becomes the number of points equidistant from the center of the Earth. Looking from the Earth's surface, such a figure forms a number of closed lines positioned at some exact altitude from the surface. Those lines with different altitudes represent examples of points with equal S_g (at equal altitudes) and different S-Outlines themselves (at different altitudes). The figure shows that graphically.

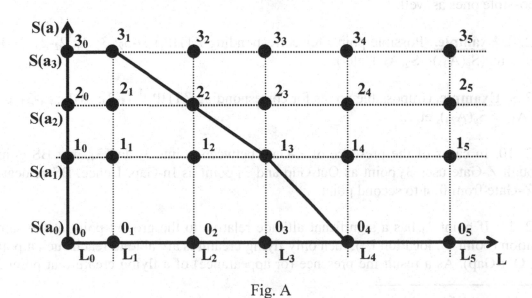

Fig. A

127

Figure [A] shows two axes, L (location according to Earth surface) and S(a) (relation between S_g and altitude above Earth surface). Axis L represents a fragment of S-Outline projection that involves the Earth surface. Under real circumstances that line is usually a curved one (in projection to Earth's surface) according to the shape of this or that S-Outline.

13.2. 3. As we can see, there are a lot of possible numbers of S-Outlines above any point of Earth's surface. Generally that number is countless because each tiny increment of altitude produces new, independent S-Outlines.

13.2. 4. Figure [13.2. 2.A] shows four different S-Outlines with different altitudes (0, 1, 2, 3). Altitude is rising according to the number of subscript ($A_3 > A_2 > A_1 > A_0$). Unlike altitude, S_g falls with rising altitude [s.3.2. 3.a] (distance from earth center; value r). Hence, $S(a_3) < S(a_2) < S(a_1) < S(a_0)$. The level with zero altitude represents an altitude of sea level.

13.2. 5. The points mentioned as numbers with subscripts are the points crossing a vertical line and a line of exact altitude. For example, point 2_3 means the point crossing altitude A_2 and a vertical line from location L_3 at Earth's surface.

13.2. 6. Line 3_0-3_1-2_2-1_3-0_4-0_5 represents an example of Earth's surface. Any point positioned below that line is a ground point, and above it is an air point. For example, 1_2 is a ground (underground) point; 1_3 is an air-ground point, where Earth's surface exists as well as air; and 1_4 is an air point.

13.2. 7. There are a lot of possible Z-Trajectories according to the points mentioned in figure [13.2. 2.A] because the S-Outline contains a countless number of ZTHP that can be used as first and last point (In-Gap and Out-Gap) of Z-Trajectory [r.7.1. 34]. The following examples show some number of possible ZTHP locations, and some impossible ones as well.

13.2. 8. **Example.** (Possible points for corresponding ZTHP.) 3_1-3_2; 3_3-3_4; 2_2-2_5; 1_2-1_4; 0_1-0_5; etc. ($S_g(A_{31})$= $S_g(A_{32})$, etc.).

13.2. 9. **Example.** (Impossible points for corresponding ZTHP.)[1] 1_1-2_4; 3_2-1_5; 2_1-3_4; etc. ($S_g(A_{11})$≠ $S_g(A_{24})$, etc.).

13.2. 10. In cases of the presence of Z-Gate between points 3_3 and 3_4, any BS going through Z-Gate uses 3_3 point as Out-Gap and 3_4 point as In-Gap. Hence, BS relocates by Z-Gate from first to second point.

13.2. 11. If point 3_3 has a significant altitude relative to the ground point at the same location (point 1_3, location L_3), then only flying creatures are able to reach such a point (3_3, Out-Gap). As a result the presence (or appearance) of a flying creature at point 3_4

---

[1] Law of conservation restriction.

(or around that point) would be unnoticeable for IB-humans. That is a real event from a human's point of view because both points are reachable for such a creature by RWT.

13.2. 12. In cases of the presence of Z-Gate between points 0_4 and 0_5, any BS going through Z-Gate uses 0_4 point as Out-Gap and 0_5 point as In-Gap. Hence, BS relocates by Z-Gate from first to second point (as well as in case of [13.2. 10]). That type of relocation looks like the previous one from a human's point of view because both points are reachable by RWT for any type of creature.

13.2. 13. In such cases any creature that is able to move on Earth's surface is relocated by the Z-Gate from point 0_4 to point 0_5. Such a relocation would go unnoticeable to IB-humans (because that is a usual event from humans' point of view).

13.2. 14. In cases of the presence of Z-Gate between points 2_2 and 2_3, any BS going through Z-Gate uses 2_2 point as Out-Gap and 2_3 point as In-Gap. Hence, BS is relocated by Z-Gate from first to second point (as well as in case of [13.2. 10]).

13.2. 15. The unique aspect from the last case [13.2. 14] is that the type of points connected by Z-Gate is unequal. First point is an air-ground point, and the last one is an air point. Hence, any creature that goes through Z-Gate finds itself at the last point of Z-Trajectory at some altitude above the Earth's surface. That altitude is equal to the difference of altitudes A_2 and A_1. Displacement itself appears from location L_2 to L_3 (from an IB's point of view).

13.2. 16. After In-Gap point, a creature establishes full interaction with the surrounding world (especially when it leaves Gap-Hood) and does everything as it would normally do in a gravity field of the Earth. Of course it falls down because of the absence of any support for the creature at In-Gap point (2_3 is air point).

13.2. 17. Hence, the scene of a falling animal from the sky will be noticed by humans (IB) and make them nonplus. That must be a noticeable event because it is impossible. from a human's point of view, to see a cow mooing and falling from the sky.

13.2. 18. In cases of the presence of Z-Gate between points 2_1 and 2_3, any BS going through Z-Gate uses 2_1 point as Out-Gap and 2_3 point as In-Gap. Hence, BS is relocated by Z-Gate from first to second point (as well as in case of [13.2. 10]).

13.2. 19. In such a case [13.2. 18], the Out-Gap appears below ground level of point L_1. The altitude difference between ground level and the level of Out-Gap (at location L_1) is equal to $A_3 - A_2$. That is a possible case for the presence Out-Gap at some depth of a river stream (for example). What will happen in that case?

13.2. 20. A school of fish goes inside a river stream, swimming down the current. When that school reaches Out-Gap, all the fish are relocated to In-Gap (2_3 point). After In-Gap point, the fish establish full interaction with the surrounding world (especially when they leave Gap-Hood) and do everything as they would in a gravity field of the

Earth. Those fish would fall down because of the absence of any support for them at In-Gap point (2_3 is air point).

13.2. 21. Hence, the scene of falling fish from the sky would turn human heads (IB). That must be a very unusual event because it is impossible, from a human's point of view, to see a fishes falling from the sky.[1]

13.2. 22. Moreover, the presence of Z-Gate gives us an answer for the usual questions for such events. "Why do only fish fall down from the sky? Why are there no rocks or tree trunks and other things falling down with the fish at the same time?"

13.2. 23. The answer is this. ZT-Time can be calculable only for the objects (BS) that move relative to the Out-Gate with enough speed that helps them go through Out-Gate before it is closed. Motionless rocks that lie on the river bed have a ZT-Time equal to infinity and are unable to use Z-Gate; this applies to any other motionless object [r.12.8. 10.b].

13.2. 24. In cases when Z-Gate appears as a zero-length relocation [s.10.2. 13], IB has no chance to separate observation from the fish that use Out-Gap at almost the same time as In-Gap, and In-Gap is separated from Out-Gap by a significant time ($n \cdot T$; [s.10.2. 12.b]). Hence, it's possible for a Z-Gate to create a significant difference between the time of corresponding Out-Gap and the In-Gap according to IB TKD.

13.2. 25. When does such a time interval becomes noticeable for the naked eyes? Obviously, it happens when a very long time between corresponding Out-Gap and In-Gap brings to the observation of the IB (human) any creatures that lived at a recent time and cannot be observed by any usual human way.

13.2. 26. "The 19th-century English naturalist Charles Darwin argued that organisms come about by evolution, and he provided a scientific explanation, essentially correct but incomplete, of how evolution occurs and why it is that organisms have features – such as wings, eyes, and kidneys – clearly structured to serve specific functions."[2]

13.2. 27. "Evolution is theory in biology postulating that the various types of plants, animals, and other living things on Earth have their origin in other preexisting types and that the distinguishable differences are due to modifications in successive generations. The theory of evolution is one of the fundamental keystones of modern biological theory."[3]

13.2. 28. "Paleontology also spelled Palaeontology, scientific study of life of the geologic past that involves the analysis of plant and animal fossils, including those of

---

[1] For additional examples, see appendix B.

[2] "Evolution." Encyclopaedia Britannica. Encyclopaedia Britannica 2008 Deluxe Edition. Chicago: Encyclopaedia Britannica, 2008.

[3] "Evolution." Encyclopaedia Britannica. Encyclopaedia Britannica 2008 Deluxe Edition. Chicago: Encyclopaedia Britannica, 2008.

microscopic size, preserved in rocks. It is concerned with all aspects of the biology of ancient life forms: their shape and structure, evolutionary patterns, taxonomic relationships with each other and with modern living species, geographic distribution, and interrelationships with the environment. Paleontology is mutually interdependent with stratigraphy and historical geology because fossils constitute a major means by which sedimentary strata are identified and correlated with one another. Its methods of investigation include that of biometry (statistical analysis applied to biology), which is designed to provide a description of the forms of organisms statistically and the expression of taxonomic relationships quantitatively.

13.2. 29. "Paleontology has played a key role in reconstructing the Earth's history and has provided much evidence to support the theory of evolution. Data from paleontological studies, moreover, have aided petroleum geologists in locating deposits of oil and natural gas. The occurrence of such fossil fuels is frequently associated with the presence of the remains of certain ancient life-forms."[1]

13.2. 30. "Paleontologists have recovered and studied the fossil remains of many thousands of organisms that lived in the past. This fossil record shows that many kinds of extinct organisms were very different in form from any now living. It also shows successions of organisms through time, manifesting their transition from one form to another."[2]

13.2. 31. In the case of [s.13.2. 25], any living creature can possibly exist (and be observed by a IB) far away from its "natural" time period. The key feature for such a possibility is ZT-Time. As well as it can be calculated for other events, it can be applied to any living creature (BS) that uses a Z-Gate. For example, it's possible to estimate ZT-Time for the pteranodon.

13.2. 32. "Pteranodon is flying reptile (pterosaur) found as fossils in North American deposits dating from about 90 million to 100 million years ago during the Late Cretaceous Period. Pteranodon had a wingspan of 7 metres (23 feet) or more, and its toothless jaws were very long and pelican-like."[3]

13.2. 33. According to the picture "Pteranodon skeleton and restoration of wings",[4] I estimate the body length of a pteranodon from head to tail to be 5 meters (in case of wingspan, it equals to 7 meters). If that creature travels in the air by a speed of 30 km/h ($30 \cdot 1000 / 3600 = 8.3$ [m/s]) (guessing velocity for calculation), ZT-Time must be equal to the following value.

---

[1] "Paleontology." Encyclopaedia Britannica. Encyclopaedia Britannica 2008 Deluxe Edition. Chicago: Encyclopaedia Britannica, 2008.

[2] "Evolution." Encyclopaedia Britannica. Encyclopaedia Britannica 2008 Deluxe Edition. Chicago: Encyclopaedia Britannica, 2008.

[3] "Pteranodon (genus Pteranodon)." Encyclopaedia Britannica. Encyclopaedia Britannica 2008 Deluxe Edition. Chicago: Encyclopaedia Britannica, 2008.

[4] "Pteranodon (genus Pteranodon)." Encyclopaedia Britannica. Encyclopaedia Britannica 2008 Deluxe Edition. Chicago: Encyclopaedia Britannica, 2008.

$$T_Z = (2 \cdot L) / V = (2 \cdot 5) / 8.3 = 1.2 \; [\text{sec}] \qquad \text{(a)}$$

13.2. 34. The value of a pteranodon's ZT-Time can be easily recalculated for any size of that creature and any speed of its travelling. The estimating calculation [13.2. 33.A] shows that ZT-Time for pteranodon appears as a few seconds according to its natural features. Hence, a pteranodon spends only seconds of its life to pass through a Z-Gate. That helps the creature to appear alive after using a Z-Gate despite a very long period of time passing by the indication of an IB's TKD (between Out-Gap and In-Gap used by a pteranodon).

13.2. 35. Which senses do an IB (human) use to see a huge, flying creature with serrated jaws and large wings hunting for its victim? Certainly the person takes an enormous impression, especially when any pteranodon-like creature (pterosaur) focuses its attention to the poor person (especially as a feeding source). Obviously, the person has a lesser impression from watching fish falling from the sky than the observation of a flying pterosaur.

13.2. 36. "Pterosaur is any of the flying reptiles that flourished during all periods (Triassic, Jurassic, and Cretaceous) of the Mesozoic Era (248 million to 65 million years ago). Although pterosaurs are not dinosaurs, both are archosaurs, or 'ruling reptiles,' a group to which birds and crocodiles also belong."[1]

13.2. 37. Here and later I refer to the phenomenon of a living creature's appearance (by using of Z-Gate) at the locations unreachable for them by RWT as the Dragon Effect. Such locations include space and time according to aspects of the particular Z-Gate that was used by a creature.

13.2. 38. "Dragon is legendary monster usually conceived as a huge, bat-winged, fire-breathing, scaly lizard or snake with a barbed tail. The belief in these creatures apparently arose without the slightest knowledge on the part of the ancients of the gigantic, prehistoric, dragon-like reptiles. In Greece the word drakon, from which the English word was derived, was used originally for any large serpent (see sea serpent), and the dragon of mythology, whatever shape it later assumed, remained essentially a snake."[2]

13.2. 39. From my point of view, statement [13.2. 38] can be used as the first noticed evidence (and written as legend) of creatures successfully passing the Z-Gate. Hence, the description of a dragon appears as mixed description from human observers, each of whom noticed different features of the prehistoric creatures like pterosaurs and dinosaurs. That leaves only the matter of the word that was used to refer to visible characteristics of such creatures (for example, bat-wings, etc.). All of them can be named as dragons as well as pterosaurs.

---

[1] "Pterosaur." Encyclopaedia Britannica. Encyclopaedia Britannica 2008 Deluxe Edition. Chicago: Encyclopaedia Britannica, 2008.

[2] "Dragon." Encyclopaedia Britannica. Encyclopaedia Britannica 2008 Deluxe Edition. Chicago: Encyclopaedia Britannica, 2008.

13.3 Z-radius and Intrusion Area

13.3. 1. Previous topics mentioned the possibility of Z-Gate to bring various creatures from one point of Earth's surface to another (from Out-Gap to In-Gap). Figure [13.2. 2.A] can be used to explain one more interesting feature related to the BS passing Z-Gates.

13.3. 2. Unlike lifeless things, living creatures have some habits and an ability to show specific behavior during any given circumstance. Despite the habits of any particular creature, all of them have some skill to keep alive, even if the circumstances are unusual.

13.3. 3. As mentioned before [r.13.1. 10], Z-Gate passage itself never produces any feeling for a living creature. Therefore, from the creature's point of view (IO in that case), its state before and after Z-Gate looks equal.

13.3. 4. For example, a bird passing Z-Gate between points 3_2 and 3_4 finds itself at In-Gap point (3_4) at its usual state. Before Z-Gate it was in the air, and after passage it still is positioned in air. That case is the best for any creature because the creature keeps its motion without any changes.

13.3. 5. In cases of Out-Gap location at point 2_2, any creature moving on the ground can keep moving only if the In-Gap appears not far from the ground. Any other case leads to the creature falling from an unfortunate altitude to the Earth's surface. That is quite dangerous for a creature because most creatures after would be damaged and unable to survive. That depends on the different creature's features and the characteristics of the landing place.

13.3. 6. If Out-Gap locates at 0_4 point and In-Gap locates at 0_5 point (both points locates at zero altitude), any creature that inhabits the water finds itself at its usual state after Z-Gate passage. Such a creature is positioned in or on water before and after Z-Gate.

13.3. 7. Any underwater creature has the same rule as a flying one because it passes through the Z-Gate from water to water (Out-Gap and In-Gap are water points). That state locates below the L axis, not shown in figure [13.2. 2.A]. That leads us to the following conclusion.

13.3. 8. *Any creature that uses Z-Gate survives undoubtedly only if the corresponding points of the gate have the same type (air to air, water to water, air-water to air-water, etc.).*

13.3. 9. Any other combination can possibly to kill the creature; at the very least, it's dangerous for the creature. For example, Z-Gate passage between points 2_2-2_4 for a cow is not dangerous. But the difference between altitudes of In-Gap (2_4) and a landing surface (point 0_4) undoubtedly damages a cow fatally if it exceeds a particular level.

13.3. 10. If fish pass Z-Gate from point 1_1 to point 1_5, they survive because fish fall down to the water (with the water surface positioned at zero altitude).

13.3. 11. Therefore, each case for each creature passing through a Z-Gate leads to different consequences for the creature. If the creature is able to leave In-Gap by itself, it moves its usual way in any direction it likes. Hence, it produces RWT after Z-Gate passage (after using of Z-Trajectory). An IB is able to observe that creature from In-Gap location (or its projection to Earth's surface) to any point that is reachable for any particular creature.

13.3. 12. Any creature can reach only some points of Earth's surface by itself and is unable to exceed them. For example any cow-like animal, after a Z-Gate passage that led it to a tiny island, can reach only the coastline of that island because a cow is unable to swim in the ocean.

13.3. 13. If a dolphin passes Z-Gate (for example, between points 0_4-0_5; air-water points) and finds itself at point 0_5 in a closed lake, it is unable to exceed the coastline of the lake because of its natural features.

13.3. 14. If any land animal (able to swim) passes Z-Gate between points 0_4 (air-ground point at zero altitude) and 0_5 (air-water point at zero altitude), it finds itself in the water (after Z-Gate passage). Because the land animal is able to swim only for a short period of time, it cannot go far from the In-Gap point (or its projection on the water's surface). After that time a creature can't survive because of exhaustion.

The examples shown in [13.3. 12–13.3. 14] lead us to the following conclusion.

13.3. 15. There is a most distant point in each particular direction reachable for each type of creature passing through a Z-Gate. Any particular type of creature is able to be detected only at points that do not exceed the maximal reachable distance from In-Gap (or its vertical projection on surface). Here and later I refer to such a distance as Z-Radius.

13.3. 16. In other words, any creature after Z-Gate passage can be detected only inside the Z-Radius. Different types of creatures have different values of Z-Radius.

13.3. 17. For example, the Z-Radius for a pteranodon that is able to fly for 24 hours without any landing with an average speed of 30 km/h is:

$$R_z = V_C \cdot T_D = 30 \cdot 24 = 720 \ [km] \tag{a}$$

In equation A, R_Z is Z-Radius, V_C is average velocity of a creature, and T_D is time of constant moving. That time is also the possible detection time. A creature is able to be detected only during that time. Obviously, the value of Z-Radius can be recalculated for any type of creature using different values for the creature's velocity and time of possible detection.

13.3. 18. Any creature is free to select the direction of its movement. Hence, any direction has an equal probability as a possible direction for a creature passing through Z-Gate. Therefore, an area that has more or less probability for the detection of such creature forms a perfect sphere across the In-Gap. Here and later I refer to such a figure as the Biological System Intrusion Possible Detection Area (BSIPDA), or simpler Intrusion Area (IA).

13.3. 19. Natural barriers can reduce that area as mentioned in [13.3. 12–13.3. 14], but no barrier can increase the Z-Radius. As a result the Z-Radius can be only reduced by a natural barrier.

13.3. 20. **Example**. Estimating the value of IA for a pteranodon can be calculated with the size of a full sphere from the Z-Radius. The following equation gives the numeric result.

$$A_I = \pi \cdot (R_Z)^2 = 3.14 \cdot 720^2 = 1{,}627{,}776 \ [km^2] \qquad\qquad (a)$$

In the equation, A_I is Intrusion Area, π is the well-known constant, and R_Z is Z-Radius [s.13.3. 17.a].

13.3. 21. In cases of a plesiosaur relocation by Z-Gate to a lake, IA becomes equal to the lake's surface area[1] because of its natural barrier (coastline of the lake). "Plesiosaur any of a group of long-necked marine reptiles found as fossils from the Late Triassic Period into the Late Cretaceous Period (215 million to 80 million years ago). Plesiosaurs had a wide distribution in European seas and around the Pacific Ocean, including Australia, North America, and Asia. Some forms known from North America and elsewhere persisted until near the end of the Cretaceous Period (65 million years ago)."[2]

13.4 In-Gap Area

13.4. 1. As mentioned above [r.13.3. 18], usually a creature can move in any direction after the Z-Gate, forming the Z-Radius and IA itself. Using the value of Z-radius for any particular type of creature, we can the estimate area of the most likely In-Gap location by the information of the same observation type from IB (observation of the same type of creature). The following figure shows that method graphically.

---

[1] If Z-Radius exceeds the length of the lake at a given direction.
[2] "Plesiosaur." Encyclopaedia Britannica. Encyclopaedia Britannica 2008 Deluxe Edition. Chicago: Encyclopaedia Britannica, 2008.

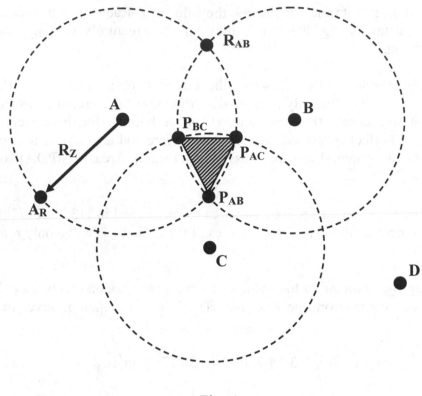

Fig. A

13.4. 2. Suppose we have three evidences from IBs located at different points around the observation of the same type of creature. Suppose that the creature was recognized by all IBs as a pteranodon. The first message about a strange creature had come from point A (a settlement, a city, etc.) at T_A point of time. Others had come from points B and C at different points of time (T_B and T_C accordingly) [s.13.4. 1.A]. That meant independent observation of different creatures, but the type of creature was the same.

13.4. 3. Putting those three points on a plain map and drawing around them three circles with the same radius (Z-Radius), we can draw a picture looking like one that is shown at [13.4. 1.A]. Each circle means the IA for some particular type of creature. Hence, any point of each circle has the same probability for presences of In-Gap where the creatures come from.

13.4. 4. The number of points belonging to any two circles simultaneously has more probability of an In-Gap presence because a creature that appears there can reach both points of observation and can be observed there. That is mentioned in figure [13.4. 1.A], as R_{AB}-P_{AC}-P_{AB}-P_{BC}-R_{AB} area. Other combinations of circles have the same areas (one area for each two circles).

13.4. 5. Among them there is only on area with points belonging to all three circles. It is mentioned in figure [13.4. 1.A] as P_{BC}-P_{AC}-P_{AB} area. From any point of that area, any point of observation is reachable by the particular type of creature. Hence, that area is the most probable one for an In-Gap location. As a result the closer to the area the IB is, the more likely for the IB to observe a strange creature

13.4. 6. Here and later I refer to the area with the most probable location of In-Gap as In-Gap Area (INGA).

13.4. 7. The Window Effect mentioned at [12.14. 17] helps to find INGA because a researcher can use the evidence of IBs from any time; it doesn't matter how distant those observations were between each other on the timeline.

13.4. 8. *If two or more different observations of the same type of creatures happen with a distance that exceeds the Z-Radius, the IBs from those points have observations of creatures that came from different INGA.*

13.4. 9. Therefore, if any point lies outside of any Intrusion Area Independent Bystanders from that point, it can never report any observation of strange creature appearances (see point D at [13.4. 1.A]).

13.5 Timeline and Fossils

13.5. 1. There is one particular physical method connected to biological systems (creatures, objects, etc.) that is related to timeline and the particular time interval where a creature lived. That method, known as radiocarbon dating, can be used to estimate a fossil's age, can find any evidence of the presence of strange creatures.

13.5. 2. "Carbon-14 dating also called radiocarbon dating, method of age determination that depends upon the decay to nitrogen of radiocarbon (carbon-14). Carbon-14 is continually formed in nature by the interaction of neutrons with nitrogen-14 in the Earth's atmosphere; the neutrons required for this reaction are produced by cosmic rays interacting with the atmosphere.

13.5. 3. "Radiocarbon present in molecules of atmospheric carbon dioxide enters the biological carbon cycle: it is absorbed from the air by green plants and then passed on to animals through the food chain. Radiocarbon decays slowly in a living organism, and the amount lost is continually replenished as long as the organism takes in air or food. Once the organism dies, however, it ceases to absorb carbon-14, so that the amount of the radiocarbon in its tissues steadily decreases. Carbon-14 has a half-life of 5,730 ± 40 years – i.e., half the amount of the radioisotope present at any given time will undergo spontaneous disintegration during the succeeding 5,730 years. Because carbon-14 decays at this constant rate, an estimate of the date at which an organism died can be made by measuring the amount of its residual radiocarbon.

13.5. 4. "The carbon-14 method was developed by the American physicist Willard F. Libby about 1946. It has proved to be a versatile technique of dating fossils and archaeological specimens from 500 to 50,000 years old. The method is widely used by

Pleistocene geologists, anthropologists, archaeologists, and investigators in related fields."[1]

13.5. 5. According to the diagram "The geologic time scale from 650 million years ago to the present, showing major evolutionary events",[2] most parts of Earth's evolution happened before that time (more than 50,000 years ago). For example, the pteranodon reptile period is dated "from about 90 million to 100 million years ago" [s.13.2. 32]. The fossils give significant information about the evolution history even if radiocarbon dating is not applicable for them.

13.5. 6. "Fossil is remnant, impression, or trace of an animal or plant of a past geologic age that has been preserved in the Earth's crust. The complex of data recorded in fossils worldwide, known as the fossil record, is the primary source of information about the history of life on Earth."[3]

13.5. 7. "Fossil organisms may provide information about the climate and environment of the site where they were deposited and preserved (e.g., certain species of coral require warm, shallow water, or certain forms of deciduous angiosperms can only grow in colder climatic conditions)."[4]

13.5. 8. Under usual circumstances, any fossil exists in some deposits according to the inhabited time of the living creature. In other words, any fossil of any particular creature is surrounded by deposits of the same geologic epoch. What if a creature passed Z-Gate and died at an unusual geologic epoch?

13.5. 9. Obviously, the fossil can be found at different geologic epochs from fossils of the same species. For example, if a Z-Gate led the pteranodon (inhabited from 90 million to 100 million years ago [s.13.2. 32]) to the tertiary period of the Cenozoic era (approximately from 18 to 60 million years ago), they fossil can be found at a period which the pteranodon never inhabited. The probability of such a discovery is very low but exceeds zero (if the probability of Z-Gate presence exceeds zero as well).

13.5. 10. Moreover, any researcher with such a discovery must give to the scientific community enough proof for the fossil location at an unusual geological era. Without such proof nobody would believe that researcher.

13.5. 11. Such a situation can be solved easily if a creature ends its travel by Z-Gate at any period that is covered by radiocarbon dating. If that method gives results for 5,000 years (for example) for a pteranodon fossil, that gives us direct proof of the last age

---

[1] "Carbon-14 dating." Encyclopaedia Britannica. Encyclopaedia Britannica 2008 Deluxe Edition. Chicago: Encyclopaedia Britannica, 2008.

[2] Encyclopaedia Britannica. Encyclopaedia Britannica 2008 Deluxe Edition. Chicago: Encyclopaedia Britannica, 2008.

[3] "Fossil." Encyclopaedia Britannica. Encyclopaedia Britannica 2008 Deluxe Edition. Chicago: Encyclopaedia Britannica, 2008.

[4] "Fossil." Encyclopaedia Britannica. Encyclopaedia Britannica 2008 Deluxe Edition. Chicago: Encyclopaedia Britannica, 2008.

where the pteranodon existed – though the creature did *not* exist and inhabit that era. Such a result would give us direct proof for the presence of Z-Gate at a geological time. Hence, such a method can be used for any fossil examination.

13.5. 12. Generally there is a low probability of finding fossils at unusual geological epochs which the original species newer inhabited. That result can be explained only by Z-Theory and gives additional proof for it.

13.5. 13. Additionally, the radiocarbon dating can be understood as a method that virtually extends our Observation Window [s.12.14. 1] from some hundred years (modern time) to a 50,000-year window with direct proof (fossils).

Chapter 14. Geographical Appearances

14.1 The Earth and Relating Objects

14.1. 1. This topic is dedicated to the explanation of Z-Gate features according to the whole planet (the Earth). According to the S_g for each point of the Earth's surface, there are few forces that produce the majority of gravity's influence. Those are the forces of gravity attraction of the Sun, the Earth, and the Moon (a particular case). The following figure shows that particular case graphically.

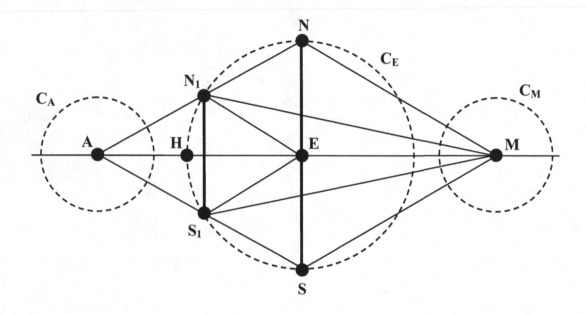

Fig. A

Figure [A] shows three celestial body centers, A, E, and M, as the centers of the Sun, the Earth, and the Moon accordingly (scale between the bodies is not accurate). The three bodies are positioned on one straight line. Hence, the figure represents the locations of the bodies at the moment of a lunar eclipse. That happens because distance AM is greater than AE. Otherwise it must be a solar eclipse because of the Moon's location between the Sun and the Earth.

14.1. 2. The Earth is positioned so that the nearest point to the Sun has the geographical coordinates 0 degrees latitude and 90 degrees longitude. As a result the Sun is located directly above that point (point H in [14.1. 1.A]). That is noon at 90 degrees longitude.

14.1. 3. According to the position of the Earth mentioned above [14.1. 2], the north and south poles have an equal distance from the Sun's center. That is shown at [14.1. 1.A] as the lines AN and AS (AN=AS). The same equality exists for the distance between

140

the poles and the Moon's center (MN=MS). Moreover, the distance between the Earth's center and the pole points are equal (EN=ES).

14.1. 4. Therefore, the pole points are equidistant from the center of each celestial body. As a result the SGFs of each celestial body at those points are equal to each other. Hence, S_g at the north and south poles is equal.

14.1. 5. We have the same result for any number of points that are equidistant from each celestial body. According to the Earth's surface, the number of such points gives us a line of zero meridian.

14.1. 6. In that case the zero meridian becomes the S-Outline [b.7.1. 27], and Shield-Gaps have a possibility to appear at the same altitude above any point of Earth's surface at that meridian. Different altitudes allow the possibility of different S-Outlines. Each system (object, thing, etc) that uses any particular S-Outline Out-Gap must use an In-Gap at the same S-Outline [s.13.2. 2.A]. Moreover, any system can use any point of S-Outline as Out-Gap or In-Gap [s.7.1. 27].

14.1. 7. Suppose an Out-Gap is located at ground level (zero meridian) in the Sahara Desert, in the southern part of Algeria. That region has the following description. "The Algerian Sahara may be divided roughly into two depressions of different elevation, separated from one another by a central north-south rise called the M'zab (Mzab). Each zone is covered by a vast sheet of sand dunes called an erg. The Great Eastern Erg (Grand Erg Oriental) and the Great Western Erg (Grand Erg Occidental), which average 1,300 to 2,000 feet (400 to 600 metres) in height, decline in elevation northward from the foot of the Ahaggar (Hoggar) Mountains to below sea level in places south of the Aures Mountains. The Ahaggar Mountains in the southern Sahara rise to majestic summits; the tallest, Mount Tahat, reaches an elevation of 9,573 feet (2,918 metres) and is the highest peak in the country."[1]

14.1. 8. Suppose an In-Gap is located far north from the previously mentioned location (Algerian Sahara) at the latitude of 51.5 degrees north (the latitude of London, UK, at zero longitude). That area has the following description according to the ground altitude above sea level. "The metropolis grew and spilled over a more or less symmetrical valley site defined by shallow gravel and clay ridges rising to about 450 feet (140 metres) on the north at Hampstead and about 380 feet (115 metres) at Upper Norwood 11 miles (18 km) to the south. Between these broken heights to the north and south, the ground falls away in a series of graded plateaus formed by gravel terraces – some at 100–150 feet (30–45 metres; the Boyn terraces, such as Islington, Putney, and Richmond) and a second and more extensive level, the Taplow terraces, at 50–100 feet (15–30 metres), on which sit the City of London, the West End, the East End, and the elevated southern districts such as Peckham, Battersea, and Clapham. The lowest ground, just a few feet above high-tide level, is the extensive floodplain of the valley

---

[1] "Algeria." Encyclopaedia Britannica. Encyclopaedia Britannica 2008 Deluxe Edition. Chicago: Encyclopaedia Britannica, 2008.

141

floor. The Thames scours the confining terraces to the north and south as it meanders toward the sea."[1]

14.1. 9. Hence, the difference in altitude at ground level between the Sahara at southern Algeria and the city of London equals approximately 500 [m] - 20 [m] = 480 [m]. Therefore, in such a case and with the presence of Spin Effect [s.12.9. 11], sand from the Sahara appears in the London sky as a sandstorm drops as much sand as the Out-Gap in Sahara takes in. ZT-Time for each grain is very little that way because of the Spin Effect and the tiny size of grains [s.12.8. 10.b].

14.1. 10. A more interesting effect appears in cases when the Out-Gap is located at some place with a plateau filled by rocks. In that case the solid underlying rock is unmovable by Out-Gap because of the infinite ZT-Time needed to pass a solid thing through an Out-Gap [r.13.2. 23]. For simple rocks, ZT-Time becomes much less and is acceptable for Z-Process. For example, the ZT-Time for a rock with a maximum size of 20 [cm] (0.2 [m]) has a ZT-Time with Spin Effect (relative velocity between the rock and Out-Gap equals 463.6 [m/s] [s.12.9. 6.a] in cases of the rock positioned not far from the equator) equal to the following.

$$T_Z = (2 \cdot L) / V = (2 \cdot 0.2) / 463.6 = 0.00086 \ [sec] \approx 0.9 \ [millisecond] \qquad (a)$$

As we can see, that must be a very rapid process. Obviously, ZT-Time can be recalculated for any size of a rock and its latitude. Latitude plays a significant role in Spin Effect because it varies its relative speed from a maximum (463.6 m/s) at zero latitude to zero at 90 latitude (north and south poles).

14.1. 11. In such a case [14.1. 10], London inhabitants would be surprised with rocks falling from the sky right onto the heads of the amazed crowd. Therefore, the type of things falling from the sky depends only on the location of the Out-Gap in space and time and has no relation to the In-Gap position. *That is the primary matter that cannot be solved only by In-Gap phenomenon observation.*

14.1. 12. The same process creates the possibility of the presence of African animals (or "strange animals") far away from their native continent. In cases when Out-Gap crosses the location of an animal, and the altitude difference between the Earth surface at the points of Out-Gap and In-Gap locations is acceptable for the creature to survive after falling from the In-Gap, the creature can be found inside the Intrusion Area around In-Gap. That is applicable for any creature despite its nature and type.

14.1. 13. In cases of a significant number of creatures that are immobile at Earth's surface (fish, for example), the area covered by those things gives us a direct projection of the In-Gap shape to Earth's surface at the In-Gap location area. If the In-Gap moves relative to the Earth's surface, it leaves more or less a wide strip of the objects that have passed through the Z-Gate successfully.

---

[1] "London." Encyclopaedia Britannica. Encyclopaedia Britannica 2008 Deluxe Edition. Chicago: Encyclopaedia Britannica, 2008.

14.1. 14. Different S-Outlines are possible for examining cases. One example is shown at [14.1. 1.A] as the N_1-S_1 line. That is a projection of an S-Outline circle on a figure plane (as well as the N-S line examined before). That line forms a circle closer to point H. As a result the radius of that circle becomes less than the radius of a maximal S-Outline (N-S) going through the north and south poles.

14.1. 15. Hence, the relocation of any system by ZTHP located at that S-Outline is possible only in the same S-Outline. Suppose that S-Outline is located so that its circle crosses the equator at points with longitude 80 West and 100 West. The same circle crosses the 90 west meridian at a latitude of 10 degrees north and south. Therefore, the points N_1 and S_1 [14.1. 1.A] have geographical coordinates as 10 degrees north and 90 degrees west (N_1), and 10 degrees south and 90 degree west (S_1). That coincides with the eastern area of the Pacific Ocean, forming a circle with a radius of about 600 miles, around the Galapagos Islands.

14.1. 16. The relation between degree and distance between points on the Earth's surface lies in the definition of a mile. "A nautical mile was originally defined as the length on the Earth's surface of one minute (1/60 of a degree) of arc along a meridian (north-south line of longitude). Because of a slight flattening of the Earth in polar latitudes, however, the measurement of a nautical mile increases slightly toward the poles. For many years the British nautical mile, or admiralty mile, was set at 6,080 feet (1.85318 km), while the U.S. nautical mile was set at 6,080.20 feet (1.85324 km). In 1929 the nautical mile was redefined as exactly 1.852 km (about 6,076.11549 feet or 1.1508 statute miles) at an international conference held in Monaco, although the United States did not change over to the new international nautical mile until 1954. The measure remains in universal use in both marine and air transportation. The knot is one nautical mile per hour."[1]

14.1. 17. Hence, according to the relation between a mile and a degree, a 20-degree separation between points N_1 and S_1 [s.14.1. 1.A] equals the following distance.

$$S = D \cdot 60 \cdot M = 20 \cdot 60 \cdot 1.852 = 2,222.4 \text{ [km]} (1200 \text{ [miles]}) \qquad (a)$$

In the equation, S is distance, D is degree angle, and M is mile length in kilometers.

14.1. 18. If Out-Gap and In-Gap are located at N_1 and S_1 points, then any ship passing N_1 point would be transferred to S_1 point by Z-Process. According to [12.9. 7] that process takes little time for any onboard IO. In the case of a latitude of 10 degree, the exact time would be a bit longer but not significantly because the relative velocity between Out-Gap and the ship becomes a bit more at the equator.

14.1. 19. As a result such a ship travels from the northern hemisphere to the southern one in case of a Z-Process described at [14.1. 18], finding itself positioned 1200 miles

---

[1] "Mile." Encyclopaedia Britannica. <u>Encyclopaedia Britannica 2008 Deluxe Edition</u>. Chicago: Encyclopaedia Britannica, 2008.

south from its previous location. That leads to a significant shrinking of the ship's route and traveling time. Hence, that ship reaches the last point of its route (port of destination) earlier than expected. That happens because the Z-Trajectory replaces part of RWT and shrinks the whole length of RWT of a ship.

14.1. 20. The same process gives an answer to the question about the absence of any evidence that mentions "ships falling from the sky". At sea level the S_g on an Out-Gap point must be equal to the same value of S_g at an In-Gap (law of conservation) [s.5.1. 10.a]. Therefore, a ship that got to an Out-Gap at sea level must come back by In-Gap *only at sea level*.

14.1. 21. The same state leads to a "navigational mistake" for the ship navigator. If the navigator locates the ship position both before Out-Gap and after In-Gap, the result of the calculations according to those positions and the time between them will confuse the navigator.

14.1. 22. **Example**. Suppose the navigator locates the first position of the ship at any given point of time. One hour later the ship successfully passed a Z-Gate. After that hour the navigator locates the ship position a second time. The result reveals to the person a relocation of 1200 nautical miles (that is the minimal distance estimation because a watercraft covers a bit more distance moving by itself for any given period of time) in 2 hours, and the average speed of the ship is 600 miles per hour (impossible for a modern watercraft). (This happens despite both locations being determined correctly.)

14.2 Consequences of Exact Equality

14.2. 1. Figure [13.2. 2.A] shows the relationship between a number of S-Outlines and a number of ZTHP located at each S-Outline. In cases of equality, S_g at each S-Outline ZTHP is located according to the description given at [13.2]. That is the usual state for any S-Outline surrounding any celestial body with any given mass. But in the real word the mass of the body can be slightly changed by time. A primary source for such changes is meteorites.

14.2. 2. "Meteorite is any fairly small natural object from interplanetary space – i.e., a meteoroid – that survives its passage through Earth's atmosphere and lands on the surface. In modern usage the term is broadly applied to similar objects that land on the surface of other comparatively large bodies. For instance, meteorite fragments have been found in samples returned from the Moon, and at least one meteorite has been identified on the surface of Mars by the robotic rover Opportunity. The largest meteorite that has been identified on Earth was found in 1920 in Namibia. Named the Hoba meteorite, it measures 2.7 metres (9 feet) across, is estimated to weigh nearly 60 tons, and is made of an alloy of iron and nickel. The smallest meteorites, called micrometeorites, range in size from a few hundred micrometres (мм) to as small as about 10 мм and come from the population of tiny particles that fill interplanetary space (see interplanetary dust particle)."[1]

14.2. 3. The next source that slowly increases planetary mass is interplanetary dust. "Interplanetary dust particle (IDP) also called micrometeoroid, micrometeorite, or cosmic dust particle a small grain, generally less than a few hundred micrometres in size and composed of silicate minerals and glassy nodules but sometimes including sulfides, metals, other minerals, and carbonaceous material, in orbit around the Sun. The existence of interplanetary dust particles was first deduced from observations of zodiacal light, a glowing band visible in the night sky that comprises sunlight scattered by the dust. Spacecraft have detected these particles as far out in space nearly as the orbit of Uranus, which indicates that the entire solar system is immersed in a disk of dust, centred on the ecliptic plane."[2]

14.2. 4. Both processes lead to a slow process of earth mass increasing. The more time that passes, the more mass that is added to the Earth. That process can be a very slow one because of the huge difference between the masses of meteorites and the Earth. But for ages the mass difference can become noticeable for measuring instruments. Moreover, the mass changing is always positive because the planet absorbs meteorites and dust from open space, acting as huge matter collector (mass collector). Here and later I refer that effect as Natural Mass Changing Effect (NMCE).

14.2. 5. As a result figure [13.2. 2.A] becomes applicable only for Z-Processes that occur between very close points of time, when the S_g difference between different points is equal to zero. In other words, the Dragon Effect described at [13.2] is impossible, taking into account NMCE in time displacement of moving system.

14.2. 6. That is true if a planet always keeps an equal value of S_g at any point of its surface. Another very interesting question appears here. What happens if the S_g has little disturbance across the planet's surface? The following figures show that case graphically.

---

[1] "Meteorite." Encyclopaedia Britannica. Encyclopaedia Britannica 2008 Deluxe Edition. Chicago: Encyclopaedia Britannica, 2008.

[2] "Interplanetary dust particle (IDP)." Encyclopaedia Britannica. Encyclopaedia Britannica 2008 Deluxe Edition. Chicago: Encyclopaedia Britannica, 2008.

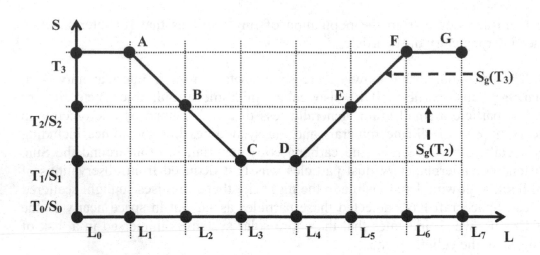

Fig. A

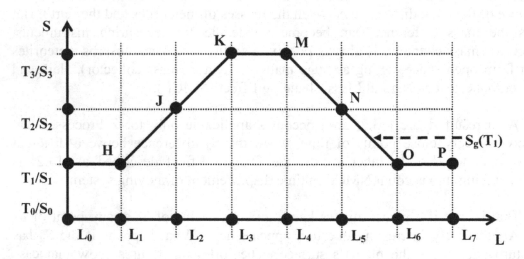

Fig. B

Both figures show similar pictures with only one difference. They represent part of the meridian. like the N-S meridian shown at [14.1. 1.A]. by the number of different locations marked by L letters with subscripts (L_0, L_1, … L_7). Those locations have the same altitude above Earth's surface and equal zero.

14.2. 7. The vertical axis shows S_g (strength of gravity force) at a given locations and at given points of time. For example S_0 exists as equal SGF for any location from L_0 to L_7 at T_0 moment of time.

14.2. 8. A lot of time passes, and S_g becomes equal to S_1 (slightly different from S_0) at the same locations and at the same altitude (zero altitude), and time reaches T_1 moment of time. Using the same method, time reaches T_2 moment of time, and S_g becomes equal to S_2. At T_3 point of time any location has S_3 value of S_g.

14.2. 9. That is correct only for the part of the N-S meridian [s. 14.1. 1.A] that covers areas with equal value of S_g. In other words that is mathematically correct.

14.2. 10. Now imagine some area that has little difference from the average value of S_g. Suppose that area lies between meridian points L_1 and L_6. Those are edge points where the difference begins and ends. Further explanations relate to figure [14.2. 6.A], until I mention the next figure.

14.2. 11. The gravity disturbance between points L_1 and L_6 at T_3 moment of time is shown graphically as a solid line, S_3-A-B-C-D-E-F-G. According to figure [14.2. 6.A], the SGF at location L_0 and at T_3 time point equals the S_3 value. That can be written as the following: $S_g(L_0,T_3)=S_3$. At point A we have this equation: $S_g(L_1,T_3)=S_3$. That is the same value of S_g. As a result a solid horizontal line connecting points S_3 and A shows the equality of SGF at locations L_0 and L_1 at the same time (T_3).

14.2. 12. At point L_2 the S_g drops its value to S_2 (point B in figure). Hence, we can write this equation: $S_g(L_2,T_3)=S_2$. If Out-Gap exists at L_0 point at T_3 time, its corresponding In-Gap can be located only at the same S-Outline – that is, points A, F, and G (according to the figure). But it cannot be located at point B because $S_g(L_0,T_3)=S_3$; $S_g(L_2,T_3)=S_2$; $S_3>S_2$. As a result, $S_g(L_0,T_3)\neq S_g(L_2,T_3)$, and Z-Trajectory between those points at T_3 point of time is impossible [s.5.1. 10.a].

14.2. 13. Which point has an equal value of S_g for B point? According to the figure that is the point with the same location but a different time, $S_g(L_2,T_2)$, when S_g was equal to S_2 value. The same points exist at E location. Those are $S_g(L_5,T_2)$ and $S_g(L_5,T_3)$ (points coincide in figure). We can write those relations in the following equation.

$$S_g(L_2,T_3) = S_g(L_2,T_2) = S_g(L_5,T_2) = S_g(L_5,T_3) = S_2 \qquad \text{(a)}$$

14.2. 14. Equation [14.2. 13.a] gives us four corresponding points with equal S_g instead of two ZTHP for the usual S-Outline in cases of a static gravity field. Each pair's points can be used for corresponding locations of Out-Gap and In-Gap. The whole number of possible Z-Trajectories that exist that way are shown in the following table.

Table A

Number	Corresponding ZTHP	Length transposition	Time transposition
1	$S_g(L_2,T_2) - S_g(L_5,T_2)$	$L_5 - L_2$	$T_2 - T_2 (0)$
2	$S_g(L_2,T_2) - S_g(L_2,T_3)$	$L_2 - L_2 (0)$	$T_3 - T_2$
3	$S_g(L_5,T_2) - S_g(L_5,T_3)$	$L_5 - L_5 (0)$	$T_3 - T_2$
4	$S_g(L_5,T_2) - S_g(L_2,T_2)$	$L_5 - L_2$	$T_2 - T_2 (0)$
5	$S_g(L_2,T_2) - S_g(L_5,T_3)$	$L_5 - L_2$	$T_3 - T_2$
6	$S_g(L_2,T_3) - S_g(L_5,T_2)$	$L_5 - L_2$	$T_3 - T_2$
7	$S_g(L_5,T_3) - S_g(L_2,T_2)$	$L_5 - L_2$	$T_3 - T_2$
8	$S_g(L_5,T_2) - S_g(L_2,T_3)$	$L_5 - L_2$	$T_3 - T_2$

14.2. 15. As we can see from [14.2. 14.A], there are four possible Z-Trajectories with zero aspect transposition (length or time), and four Z-Trajectories with transpositions that involve length and time simultaneously (numbers 5–8).

14.2. 16. Zero time transposition (Length Transposition) means simultaneous existence for Out-Gap and In-Gap according to the readings of IB's TKD. A moving system that uses those corresponding Shied-Gaps spends more or less time to Z-Process according to its dimensions. As one might remember, that is ZT-Time itself.

14.2. 17. Zero length transposition (Time Transposition) means a location of corresponding Shied-Gaps at the same point in space but not in time according to the coordinates system bound to the Earth (a celestial body). Such a relocation seems from the IB point of view as a disappearance of the observing system at a particular point of space, and an appearance of the same system at the same point of space but a different point of time [r.1.1].

14.2. 18. Other types of transposition that have different locations of Shied-Gaps according to the coordinate system bound to the Earth (a celestial body) are Space-Time Transpositions (STT). Generally any type of transposition is a Space-Time Transposition. Particular cases are possible only if the coordinate system is bound to a particular celestial body, and the IB associates itself with that body.

14.2. 19. Now suppose a system moves along the L axis [14.2. 6.A], and the time equals T_3. At L_0 location it was surrounded by S_3 level of S_g. Moving further it reaches the L_1 location. That is an edge location where S_g begins to change (but it still has value equal to S_3 at L_1 point).

14.2. 20. After some time the system reaches the L_2 location. That point has a value of S_g equal to S_2 at the moment of the system located there (point B). Moving further, the system reaches consequently points L_3 and others until L_7. According to the relocation of the system, the value of S_g changes and reaches its minimum between location L_3 and L_4 (points C and D of the figure), and restores its maximum again at location L_6.

14.2. 21. Now let's image a system that followed the same path (L_0-L_7) in some distant past (T_2), when S_g was equal to the average S_g at zero altitude level of the Earth (S_2 level). The same altitude is used by the previous system, as mentioned before. When the system (at T_2 time) reached location L_1, the value of S_g at point (L_1,T_2) equals S_2. Because $S_g(L_1,T_2) \neq S_g(L_1,T_3)$, Z-Process is impossible between those points (between location L_1 and different time points T_2 and T_3).

14.2. 22. When the system reaches location L_2, it is surrounded by the same value of S_g equal to S_2. But that point has the same level of S_g at T_3 time [s.14.2. 6.A]. That is point B in the figure. As a result, $S_g(L_2,T_2)=S_g(L_2,T_3)$. That gives those points the possibility to be a ZTHP. In other words those locations can be points of corresponding Shield-Gaps [r.5.1. 10.a]. That leads to the possibility for a moving system to use Z-Process to change its location from point (L_2,T_2) to (L_2,T_3) (time transposition because $T_3 \neq T_2$). Moving further, the system matches the same physical processes as well as before the Z-Gate, and the state of the system is the same as before Z-Gate. Hence, the behavior of a same type of system that uses T_3 time to reach point B and T_2 time with Z-Process to reach same point (point B, (L_2,T_3) as well) are identical after that point.

14.2. 23. That happened if the moving system reaches point (L_2,T_2) before (L_2,T_3), according to IO's TKD (the observer who travels with the system and is involved in the Z-Process as part of the system). Otherwise if the moving system reaches point (L_2,T_3) before (L_2,T_2), Z-Process caused its transposition to the corresponding Shield-Gap point of Z-Trajectory (L_2,T_2) because $S_g(L_2,T_2)=S_g(L_2,T_3)$. Hence, for such a system the Shield-Gap points have the same location but different sequences of use. That means the names for Out-Gap and In-Gap are relative for a system using Z-Gate. Z-Trajectory appears as connections between two ZTHP regardless of the transposition direction. Therefore, a system moving from T_2 time to T_3 time has the same Shield-Gaps as well as the system moving in opposite direction. In other words, the Out-Gap for one system becomes the In-Gap for a system moving in the opposite direction.

14.2. 24. Hence, the system using Z-Gate in direction (L_2,T_3) - (L_2,T_2) appears for IB at T_3 time as something that "disappeared without a trace". For IB located at T_2 time point, the same system using the same Z-Gate looks like a system that "strangely appeared from nowhere".

14.2. 25. Moreover, the IB at time T_3 notices a strange relationship between the phenomenon and area at the Earth's surface. That is true because such a phenomenon can be observable (and possible to exist) only between locations L_1 and L_6 (according to [s.14.2. 6.A]) where a gravity anomaly has taken place.

14.2. 26. Additionally, from the IB's point of view, who is located at T_2 time equal to S_g between point L_2 and any other point on the L axis (for example $S_g(L_2,T_2)=S_g(L_6,T_2)$), it leads to the possibility of a ZTHP presence at any point of the same S-Outline belonging to the same time (T_2). As a result the IB at T_2 time has no idea of the strange thing's origin, which is observable from time to time and is coming from the S-Outline (there $S_g=S_2$ and $T=T_2$; T is observation time).

14.2. 27. Moreover, in other cases, anything that uses Z-Trajectory from $(S_2;T_2)$ S-Outline is able to use In-Gap at points B or E at T_3 time. Therefore, from the IB point of view that is located at T_3 time, strange creatures come from strange points L_2 and L_5.

14.2. 28. The same process is applicable for any point between L_1 and L_6 with additional consequences. The larger the distance between the L_1 location and the particular point, the more possible time difference between ZTHP. For example, location L_3 has time difference for T_3-T_1 in the case of (L_3,T_1) - (L_3,T_3) transposition, which is more than the (L_2,T_2) - (L_2,T_3) time difference according to the IB's TKD.

14.2. 29. Also, anything using Z-Gate between time T_1 and T_3 has a chance to use In-Gap between locations L_1 and L_6. Therefore, that area has a higher probability for strange things to appear that used the Out-Gaps from T_1 time to T_3 time. All those things come from the past.

14.2. 30. *Hence, a gravity anomaly with decreased S_g allows, and increases the probability for, In-Gap presence with a corresponding Shield-Gap located in the past.* Here and later I refer that as the Negative Time Shift effect (NTSE).

14.2. 31. A gravity anomaly with a different sign has different properties and leads to different possibilities for Z-Process. Such an anomaly that has more value for S_g than the planet average is shown in figure [14.2. 6.B]. Letters and other marks in that figure are equal to figure [14.2. 6.B] with only one difference: the value of S_g between locations L_1 and L_6 is higher than average (S_1); the current time is T_1. The following explanations relate only to figure [14.2. 6.B].

14.2. 32. Like the previous case, with a difference in S_g for the same locations, it leads to the possibility of the presence of ZTHP not only at time T_1 (current time for the figure). According to [14.2. 6.B], a possible Z-Trajectory appears between points (L_3, T_1) and (L_3, T_3). That possibility connects the S-Outline with value S_3 located at time T_3 to an area between locations L_3 and L_4 (time T_1). Hence, the probability of the presence for anything using Z-Gate with Out-Gap located at that S-Outline and In-Gap between locations L_3 and L_4 at different time is significantly higher than at other locations. Generally only that area (between L_3 and L_4) is possible for transposition from T_3 time.

14.2. 33. From the IB point of view located at time T_1, the area between L_3 and L_4 is the area that brought up very strange things from nowhere. But that nowhere connects to some future time that differs from the time of IB by a time interval between T_1 and T_3. As a result the IB can see at that area the appearance of, say, a strange apparatus instead of a prehistoric creatures (as in the case of NTSE). Any other rule for transpositions between points are equal for the case of figures [14.2. 6.A] and [14.2. 6.B].

14.2. 34. *Hence, the gravity anomaly with an increased S_g allows and increases the probability for In-Gap presence with a corresponding Shield-Gap located in the future.* Here and later I refer to that as the Positive Time Shift effect (PTSE).

14.2. 35. It's time to see appendix C, which shows a map of "The variation in the gravitational field, given in milliGals (mGal), over the Earth's surface gives rise to an imaginary surface known as the geoid. The geoid expresses the height of an imaginary global ocean not subject to tides, currents, or winds. Such an ocean would vary by up to 200 metres (650 feet) in height because of regional variations in gravitation".[1] According to the gravitation, we have some additional notion that I put below mostly for the readers who are not familiar with that area of knowledge.

---

[1] Encyclopaedia Britannica, Inc.

14.2. 36. "The acceleration g varies by about 1/2 of 1 percent with position on the Earth's surface, from about 9.78 metres per second per second at the Equator to approximately 9.83 metres per second per second at the poles. In addition to this broad-scale variation, local variations of a few parts in 10^6 or smaller are caused by variations in the density of the Earth's crust as well as height above sea level."[1]

14.2. 37. "The gravitational potential at the surface of the Earth is due mainly to the mass and rotation of the Earth, but there are also small contributions from the distant Sun and Moon. As the Earth rotates, those small contributions at any one place vary with time, and so the local value of g varies slightly. Those are the diurnal and semidiurnal tidal variations. For most purposes it is necessary to know only the variation of gravity with time at a fixed place or the changes of gravity from place to place; then the tidal variation can be removed. Accordingly, almost all gravity measurements are relative measurements of the differences from place to place or from time to time."[2]

14.2. 38. This definition describes the unit of gravity. "Because gravity changes are far less than 1 metre per second per second, it is convenient to have a smaller unit for relative measurements. The gal (named after Galileo) has been adopted for this purpose; a gal is one-hundredth metre per second per second. The unit most commonly used is the milligal, which equals 10^{-5} metre per second per second – i.e., about one-millionth of the average value of g."[3]

14.2. 39. "Modern gravimeters may have sensitivities better than 0.005 milligal, the standard deviation of observations in exploration surveys being of the order of 0.01–0.02 milligal."[4]

14.2. 40. "The value of gravity measured at the terrestrial surface is the result of a combination of factors:

1. The gravitational attraction of the Earth as a whole
2. Centrifugal force caused by the Earth's rotation
3. Elevation
4. Unbalanced attractions caused by surface topography
5. Tidal variations
6. Unbalanced attractions caused by irregularities in underground density distributions"[5]

[1] "Gravitation." Encyclopaedia Britannica. Encyclopaedia Britannica 2008 Deluxe Edition. Chicago: Encyclopaedia Britannica, 2008.
[2] "Gravitation." Encyclopaedia Britannica. Encyclopaedia Britannica 2008 Deluxe Edition. Chicago: Encyclopaedia Britannica, 2008.
[3] "Gravitation." Encyclopaedia Britannica. Encyclopaedia Britannica 2008 Deluxe Edition. Chicago: Encyclopaedia Britannica, 2008.
[4] "Gravitation." Encyclopaedia Britannica. Encyclopaedia Britannica 2008 Deluxe Edition. Chicago: Encyclopaedia Britannica, 2008.
[5] "Gravitation." Encyclopaedia Britannica. Encyclopaedia Britannica 2008 Deluxe Edition. Chicago: Encyclopaedia Britannica, 2008.

14.2. 41. "The mass of the Earth can be calculated from its radius and g if G is known. G was measured by the English physicist-chemist Henry Cavendish and other early experimenters, who spoke of their work as 'weighing the Earth.' The mass of the Earth is about 5.98×10^{24} kg while the mean densities of the Earth, Sun, and Moon are, respectively, 5.52, 1.43, and 3.3 times that of water."[1]

14.2. 42. Now it's possible to do some calculations according to Earth's mass and related time. As mentioned in [14.2. 41] the current mass of the Earth equals 5.98×10^{24} kg. The Earth produces an acceleration (g) [s.14.2. 36] of "9.78 metres per second per second at the Equator to approximately 9.83 metres per second per second at the poles". We can use the average value of g equal to 9.8 metres per second per second for the following calculations.

14.2. 43. That acceleration (9.8 metres per second per second) gives the value of 9.8×10^5 in milligal because one milligal equals 10^{-5} metre per second per second [s.14.2. 38]. Hence, we have the first relation between Earth's mass (5.98×10^{24} kg) and g (9.8×10^5 milligal).

14.2. 44. If the given mass of the Earth produces the exact value of g it must produce a related value with mass deviation. Moreover, the less the mass deviation that exists, the more linear relation that occurs between the mass of the planet and the value of g.

14.2. 45. Using notion [14.2. 44] and g deviation according to the map (appendix C), and the eastern U.S. coastal line (the west part of Atlantic Ocean) being equal to -50 mGal, we can calculate the mass of the Earth that produced the same value of g as the average at the planet surface. Hence, we have following proportions.

5.98×10^{24} kg $= 9.8 \times 10^5$ mGal
X-Mass $= (9.8 \times 10^5$ mGal $- 50$ mGal$) = 980,000 - 50 = 979,950$ mGal

The solution gives us this result.

X-Mass $= (5.98 \times 10^{24}$ kg$) \times 979,950 / 980,000 = 5.979695 \times 10^{24}$ kg

As a result the planetary mass deviation becomes equal to 5.98×10^{24} kg - 5.979695×10^{24} kg $= 0.000,305 \times 10^{24}$ kg (3.05×10^{20} kg). Most sources agree that the whole value of Earth's mass increasing due to meteorites and dust equals 22,000 tons per year (22×10^6 kg/year). Therefore, the time difference between the present gravity state of the Earth and the mass deviation equals $0.000,305 \times 10^{24}$ kg and has the following time result.

$$3.05 \times 10^{20} \text{ kg} / 2.2 \times 10^7 \text{ [kg/year]} = 1.386364 \times 10^{13} \text{ [years]} \qquad \text{(a)}$$

[1] "Gravitation." Encyclopaedia Britannica. Encyclopaedia Britannica 2008 Deluxe Edition. Chicago: Encyclopaedia Britannica, 2008.

Here and later I refer that time as Maximal Transposition Time (MTT). That time can be recalculated for any point of Earth's surface with any given value of g deviation.

14.2. 46. That result from [14.2. 45.a] is more than all the time of life on Earth's surface. The most resent time has a estimation of 2.5 billion years ago from the early Proterozoic Era (2.5×10^9 years ago). "Proterozoic Eon is the younger of the two divisions of Precambrian time, extending from 2.5 billion to 540 million years ago. It is often divided into the Early Proterozoic Era (2.5 to 1.6 billion years ago), the Middle Proterozoic Era (1.6 billion to 900 million years ago), and the Late Proterozoic Era (900 to 540 million years ago). Proterozoic rocks have been identified on all the continents and often constitute important sources of metallic ores, notably of iron, gold, copper, uranium, and nickel. It is thought that the many small protocontinents that had formed during early Precambrian time coalesced into one or several large landmasses by the initial segment of the Proterozoic. Rocks of the Proterozoic contain many definite traces of primitive life-forms – e.g., the fossil remains of bacteria and blue-green algae"[1]

14.2. 47. Moreover, the calculation result of [14.2. 45.a] exceeds the estimation of the minimum age of the Earth, equal to 4,600,000,000 (4.6×10^9) years. "Such techniques[2] have had an enormous impact on scientific knowledge of Earth history because precise dates can now be obtained on rocks in all orogenic (mountain) belts ranging in age from the early Archean (about 3,800,000,000 years old) to the late Tertiary (roughly 20,000,000 years old). The oldest sedimentary and volcanic rocks in the world are found at Isua in western Greenland; they have an isotopic age of approximately 3,800,000,000 years – a fact first established in 1972 by Stephen Moorbath of the University of Oxford. The oldest rocks on Earth are highly recrystallized granites (tonalitic gneisses from the Acasta River area in northwestern Canada), which Sam Bowring of the Massachusetts Institute of Technology has shown have an isotopic age of 4,030,000,000 years. Also by extrapolating backward in time to a situation when there was no lead that had been produced by radiogenic processes, a figure of about 4,600,000,000 years is obtained for the minimum age of the Earth. This figure is of the same order as ages obtained for certain meteorites and lunar rocks."[3]

14.2. 48. Additional calculations give us the multiplicity of time estimation according to the age of the Earth.

$$(1.386364 \times 10^{13}) \text{ [years]} / (4.6 \times 10^9) \text{ [years]} = 0.301 \times 10^4 = 3.01 \times 10^3 \qquad \text{(a)}$$

Hence, the age of the Earth is approximately 3,000 times less than the MTT that appears by gravitational variation around the Earth's surface. *Therefore, the Earth mass*

---

[1] "Proterozoic Eon." Encyclopaedia Britannica. Encyclopaedia Britannica 2008 Deluxe Edition. Chicago: Encyclopaedia Britannica, 2008.
[2] Using the mass spectrometer.
[3] "Earth sciences." Encyclopaedia Britannica. Encyclopaedia Britannica 2008 Deluxe Edition. Chicago: Encyclopaedia Britannica, 2008.

deviation caused by space factors (falling mass) is negligible for Z-Process and allows it to appear between any two Earth evolution time points.

14.2. 49. In other words, the smallest mass deviation that can be indicated by modern devices (0.005 mGal) [s.14.2. 39] has the following calculation for MTT (using the same method as for [14.2. 45]).

g deviation

$$(9.8 \times 10^5 \text{ [mGal]} - 0.5 \times 10^{-2} \text{ [mGal]}) = 980,000 - 0.005 = 979,999.995 \text{ [mGal]} \tag{a}$$

Past time Earth mass

$$\text{X-Mass} = (5.98 \times 10^{24} \text{ [kg]}) \times 979,999.995 \text{ [mGal]} / 980,000 \text{ [mGal]}$$
$$= 5.979999 \times 10^{24} \text{ kg} \tag{b}$$

Earth mass deviation

$$5.98 \times 10^{24} \text{ [kg]} - 5.979999 \times 10^{24} \text{ [kg]} = 0.000,001 \times 10^{24} \text{ [kg]} (1 \times 10^{18} \text{ [kg]}) \tag{c}$$

MTT value

$$1 \times 10^{18} \text{ kg} / 2.2 \times 10^7 \text{ [kg/year]} = 1.818 \times 10^{11} \text{ [years]} \tag{d}$$

Hence, even a 0.005 mGal deviation causes a Z-Process possibility between two time points separated by 1.818×10^{11} [years] according to the IB's TKD. That is (1.818×10^{11}) [years] / (4.6×10^9) [years] $= 0.4 \times 10^2 = 40$ times greater than the minimal estimation of the Earth's age $(4.6 \times 10^9$ [years]). That leads us to following conclusion.

14.2. 50. According to the deviation of S_g in time around the Earth, the mass of the planet can be used as constant.

14.2. 51. The only difference between the deviation of g and S_g is that the S_g value is independent from the planet rotation and depends only on the value of gravity force attraction between the planet and the unit mass located at a given point of space (around the Earth).

14.2. 52. Rule [14.2. 50] produces the following consequence. The whole of Earth's surface and its neighboring points above and below it can be used as ZTHP regardless of the time separated by two corresponding S-Gaps (Shied-Gaps).

14.2. 53. The only restriction for possible S-Gap points is the equality of S_g (that can be showed according to the deviation of g) at a possible S-Outline. That is fully compatible with [5.1. 10.a].

14.2. 54. Such a consideration can be shown graphically.

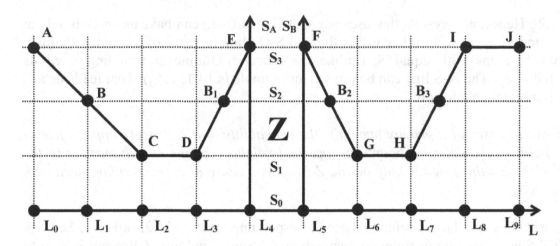

Fig. A

Figure [A] shows a number of possible S-Outlines at two different periods of time, T_A and T_B. Each S-Outline is marked by the letter S with a subscript. There are four different S-Outlines in the figure, from S_0 to S_3.

Different points of locations on the Earth's surface directly below S-Outlines are marked by the letter L with subscripts. Locations from L_0 to L_4 are placed at T_A time interval, and locations from L_5 to L_9 exist at T_B time interval. The relationship between left and right parts of the figure is possible only by Z-Process, mentioned by the capital Z between two axes of S_g marked as S_A and S_B.

14.2. 55. Broken line A-B-C-D-E [s.14.2. 54.A] represents a gravity anomaly that occurred in some distant past (T_A time interval). A different broken Line F-G-H-I-J represents a gravity anomaly at different time interval T_B. That time interval can be understood as the present day.

14.2. 56. Each S-Outline [s.14.2. 54.A] is shown as its parts between locations L_0-L_4 (T_A) and L_5-L_9 (T_B). Projections of each S-outline on Earth's surface coincides with the N-S line in figure [14.1. 1.A].

14.2. 57. Using the main rule of Z-Process (the possibility to appear between points of S-Outline) [r.7.1. 27; 7.1. 34; 9.5. 3; 11.2. 24], we can find possible corresponding points at the T_B time interval for each S-Gap that exists at the T_A time interval. Such points must occupy the same S-Outline.

14.2. 58. According to the deviation of S_g value at each gravity anomaly zone, some possible areas for ZTHP appear in figure [s.14.2. 54.A]. First of all these are the areas of L_2-L_3 location and L_6-L_7 location. That happens because S_g has an equal value (S_1) at each area. Therefore, each point between L_2-L_3 and L_6-L_7 can be used as ZTHP (corresponding points).

14.2. 59. Hence, any system that uses any Out-Gap at L_2-L_3 can have an In-Gap only at L_6-L_7, and vice versa. In other words the real S-Outline in cases of a gravity anomaly shrinks to areas with equal S_g (unlike theoretical S-Outline surrounding a whole celestial body. The N-S line can be used as an example [s.14.1. 1.A]). That leads us to a very important conclusion.

14.2. 60. *In cases of a gravity anomaly, the probability of a Z-Process appearance at each point of the celestial body's surface and neighbor points (Earth surface in the case of the Earth) depends only on the Z-Process's cause of corresponding area with an equal S_g.*

14.2. 61. Here and later I refer to these corresponding areas [s.14.2. 60] as Z-Sectors. Each Z-Sector has one or more corresponding Z-Sectors, and any Z-Process is able to appear only between Z-Sectors. In the case of a gravity anomaly absence, Z-Sectors turn into the number of simple S-Outlines surrounding the celestial body.

14.2. 62. According to figure [s.14.2. 54.A] there are four Z-Sectors, at locations L_0-L_2, L_3-L_4, L_5-L_6, and L_7-L_8. The value of S_g at those Z-Sectors varies but keeps itself between the values of S_1 and S_3. Therefore, each point between L_0-L_2 has a corresponding point at any other Z-Sector. For example point B from L_0-L_2 Z-Sector has the corresponding points B_1, B_2, and B_3 in Z-Sector of L_3-L_4, L_5-L_6, and L_7-L_8 accordingly. Those points have a value of S_g equal to S_2.

14.2. 63. As a result the probability of anything using In-Gap in Z-Sector depends only on the probability of the presence of the same thing in a corresponding Z-Sector (at Out-Gap). That notion can be used as an indirect description for any Z-Sector and can be derived from observation of a Z-Phenomenon at any particular area.

14.3 Z-Sectors and the Surface of the Earth

14.3. 1. It's time to use the map from appendix C to try to find possible locations for Z-Sectors. According to [14.2. 63] each Z-Sector must be a combination of a gravity anomaly and an area with high probability of strange events or with appearance-disappearance processes of any system. Any events mentioned in this work must have a higher probability of existence (observation) in such areas.

14.3. 2. The map presents some areas with significant positive and negative deviation of g (S_g). Negative deviation areas of S_g (NDA) are marked in blue. The more deviation that exists at a point, the darker color that marks the point. As we can see there are a few large NDA with the highest value of deviation.

North and South America

    a. The large area at the north-east part of North America

    b. Off the east coast from North and South America (the west part of Atlantic Ocean)

Eurasia

    c. The continental area that coincides partly with a high mountain region (the Himalayas)

    d. The area that includes South India and a significant part of Indian Ocean (the largest NDA)

Australia

    e. The area around the south-west part of Australia (off the coast)

Antarctica

    f. The large area located south-east from Australia (the continental part of Antarctica and off the coastal area)

Africa

    g. A little area in central Africa

    h. The area located along the east coast

14.3. 3. Generally each area mentioned above [14.3. 2] can be one of corresponding Z-Sectors. If so the Z-Sector must exhibit a tendency to demonstrate strange phenomena including the disappearance and appearance of objects.

14.3. 4. Moreover, the type of events must show some tendencies according to any particular Z-Sector. For example, suppose one of the corresponding Z-Sectors is located at an area and time that was populated by a particular type of creature. As a result the probability for that type of creature to be involved in the Z-Process is significantly higher than for any other creature type.

14.3. 5. Additionally, the type of creatures that can be involved in the Z-Process from any Z-Sector depends on the type of S-Gaps that appear there. For example, any S-Gap located in the air can be reached only by things that are able to fly (aircraft, flying animals, etc.). The same result exists for S-Gaps located at the air-water line. They can be reached by flying animals and watercrafts. Generally such a state is possible because the Z-Process itself is independent on any relocating system [r.12] (as well as any other physical process).

157

14.3. 6. *Therefore, the type of event must be the same for any Z-Sector.* If there is a probability for the "appearance from nowhere of strange flying creatures", that means there is an above-zero probability for the presence of an air-to-air Z-Gate [r. 13.2. 10–13.2. 11]. As a result there is an above-zero probability for any flying system (object) to "disappear without a trace" despite its nature (aircraft, sounding balloon, etc.) in the case of Back-Transposition.

14.3. 7. Back-Transposition means the opposite transposition by a Z-Gate. That happens when a moving system uses the same S-Gaps in a different sequence. For example, there is a system that used a Z-Gate from point L_1 to point L_2. Hence, there was an Out-Gap at L_1 and an In-Gap at L_2 for that system. A different system that uses the same locations from L_2 to L_1 but has an Out-Gap at L_2 point and an In-Gap at L_1 point does a Back-Transposition.

14.3. 8. The areas of Z-Sectors with the most famous reputation is the Bermuda Triangle. Its location coincides with part of the [14.3. 2.b] zone. Information from the map [appendix C] gives us more details. According to the deviation of g (S_g), the Bermuda Triangle is only part of a significantly larger area that occupies the north-west part of the Atlantic Ocean. Hence, any part of that area is able to show the same strange phenomena. That is Z-Sector (B).

14.3. 9. Using geological information about the era of presence of any creature that was observed there, we can collect information about the location and time where the corresponding Z-Sector was placed. For example, if most of the strange creatures observed there (Z-Sector (B)) are pteranodon-like creatures, that means that The corresponding Z-Sector had a place at some location in past that has a difference in time from the current indications of an IB's TKD, from about 90 million to 100 million years ago [r.13.2. 32]. Moreover, if only flying reptiles appear at the Z-Sector, that means there is a probability of the presence only for air-to-air Z-Gates (both S-Gaps located in the air). As a result the IB can expect strange disappearances for modern aircrafts at the same Z-Sector that uses the same type of Z-Gate (air S-Gaps).

14.3. 10. Because a Z-Sector has a corresponding one, any Back-Transposition brings the relocating system (object) to the same time and place that is (was) source of the systems (objects) for the opposite transposition. Hence, any aircraft that uses an S-Gap at the same Z-Sector is relocated to some point above Earth's surface that was inhabited by pteranodon-like creatures – and far from the present time (approximately for 90 million to 100 million years ago). Obviously, after the crash of that aircraft, we have no chance to find any evidence of that event after about 100 million years. In effect, the aircraft "disappeared without a trace" that way. Moreover, we have no evidence about such a catastrophe because it happened in the distant past, *not at the time when radio contact was broken.*

14.3. 11. Furthermore according to the map [appendix C], the nearest edge of Z-Sector (B) is located close to the east coast of North America. As a result that coast can be part of the Intrusion Area for pteranodon-like creatures according to [13.3. 18]. The probability of observing strange flying creatures here is significantly higher than any other part of the continent.

14.3. 12. The next point of Z-Sector (B) with the closest location to a continent is its south-west point where the Z-Sector (B) "touches" South America.

14.3. 13. **Z-Sector (H)**. One more very important area is Z-Sector (H) [s.14.3. 2.h]. That area is not as famous as the Bermuda Triangle (part of Z-Sector (B)), but it's more interesting according to the result of Z-Process related especially to the Intrusion Area.

14.3. 14. In that area there is a very strange fish called the coelacanth. "Coelacanth is any of the lobe-finned bony fishes of the order Crossopterygii. Members of the related but extinct suborder Rhipidistia are considered to have been the ancestors of land vertebrates. In some systems of classification, the coelacanths and rhipidistians are considered separate orders, members of the subclass Crossopterygii.

14.3. 15. "Modern coelacanths are deep-sea fishes of the family Latimeriidae. The name refers to their hollow fin spines (Greek: koilos, 'hollow'; akantha, 'spine'). The modern coelacanths are bigger than most fossil coelacanths and are powerful predators with heavy, mucilaginous bodies and highly mobile, limblike fins. They average 5 feet (1.5 m) in length and weigh about 100 pounds (45 kg).

14.3. 16. "Coelacanths appeared about 350 million years ago and were abundant over much of the world; the genus Coelacanthus has been found as fossils in rocks from the end of the Permian, 245 million years ago, to the end of the Jurassic, 144 million years ago. Coelacanthus, like other coelacanths, showed a reduction in bone ossification and a general trend toward a marine mode of life away from the earlier freshwater environment.

14.3. 17. "It was long supposed that coelacanths became extinct about 60 million years ago, but in 1938 a living member (Latimeria chalumnae) was netted in the Indian Ocean near the southern coast of Africa. Rewards were offered for more specimens, and in 1952 a second (named Malania anjouanae but probably not separable from Latimeria) was obtained from near the Comoros Islands. Several others have been caught in that area. It was later discovered that these fishes were well known to the islanders, who considered the flesh edible when dried and salted; the rough scales were used as an abrasive."[1]

14.3. 18. There are a few key phrases in the quotation given above [s.14.3. 14–14.3. 17]. These are clear evidence for the existence of Z-Sectors.

   a. The fish was netted in the Indian Ocean near the southern coast of Africa.
   b. The next fish was obtained near the Comoros Islands
   c. Coelacanths allegedly became extinct about 60 million years ago

---

[1] "Coelacanth." Encyclopaedia Britannica. Encyclopaedia Britannica 2008 Deluxe Edition. Chicago: Encyclopaedia Britannica, 2008.

d. In 1938 a living member (Latimeria chalumnae) was netted

e. Modern coelacanths are deep-sea fishes

14.3. 19. Hence, the area of the netted fish covers the Indian Ocean near the southern coast of Africa and the Comoros Islands. Looking closer at those islands, we have the following information. "The Comoros are a group of islands at the northern end of the Mozambique Channel of the Indian Ocean, between Madagascar and the southeast African mainland, about 180 miles (290 km) off the eastern coast of Africa. The islands from northwest to southeast include Ngazidja (Grande Comore), Mwali (Moheli), Nzwani (Anjouan), and Mayotte (Maore)."[1]

14.3. 20. It's easily visible on the map [appendix C] that the region is located not far from the south edge of Z-Sector (H). Also, there is an ocean current there that runs along eastern Africa's coast. "South of the monsoon region, a steady subtropical anticyclonic gyre exists, consisting of the westward-flowing South Equatorial Current between 10° and 20° S, which divides as it reaches Madagascar. One branch passes to the north of Madagascar, turns south as the Mozambique Current between Africa and Madagascar, and then becomes the strong, narrow (60 miles) Agulhas Current along South Africa before turning east and joining the Antarctic Circumpolar Current south of 45° S; the other branch turns south to the east of Madagascar and then curves back to the east as the South Indian Current at about 40° to 45° S."[2]

14.3. 21. The current crosses the south part of Z-Sector (H), and as a result any fish can be brought from that area to the area between Africa and Madagascar. And the first islands the current meets are Comoros Islands [s.14.3. 19]. Therefore, the probability to discover strange fish traveling in that current is significantly higher than the rest of that area.

14.3. 22. Moreover, the probability to discover the same type of fish exceeds zero at any point of Z-Sector (H) that includes the eastern Africa coast, from the latitude of Cape Gwardafuy (Guardafui), Somalia (approximately 12° N) to the latitude of Madagascar's North Cape (approximately 12° S), and it approximately reaches a longitude of 49° E.

14.3. 23. That is Z-Sector (H) itself, and the Intrusion Area in such a case is difficult to calculate because the creature appears in its usual environment [r.13.3. 8] and is free to move any direction, supporting itself by feeding.

14.3. 24. Other islands that are located near edge of Z-Sector (H) are the Seychelles. "Seychelles is officially Republic of Seychelles island republic in the western Indian Ocean, consisting of about 115 islands. Situated between latitude 4° and 11° S and longitude 46° and 56° E, the major islands of Seychelles are located about 1,000 miles

---

[1] "Comoros." Encyclopaedia Britannica. Encyclopaedia Britannica 2008 Deluxe Edition. Chicago: Encyclopaedia Britannica, 2008.

[2] "Indian Ocean." Encyclopaedia Britannica. Encyclopaedia Britannica 2008 Deluxe Edition. Chicago: Encyclopaedia Britannica, 2008.

(1,600 km) east of Kenya."[1] That is the most probable location on dry land for netting strange fish. Another probable location for strange fish hunting is Z-Sector (H) itself.

14.3. 25. According to [14.3. 17], "It was long supposed that coelacanths became extinct about 60 million years ago". And it was correct. Moreover, that is still correct because the presence of live coelacanths does not mean their surviving for the last 60 million years. The First reason for that is the absence of necessary natural circumstances for their survival. In other words, nobody explains the special features of the region where coelacanths inhabit today. There is also no explanation for how that species stayed unchanged for about 60 million years.

14.3. 26. If coelacanths (a simple species) are able to survive in a varying environment without any changes, it seems contrary to the modern view on evolutionary theory. "Biological evolution is a process of descent with modification. Lineages of organisms change through generations; diversity arises because the lineages that descend from common ancestors diverge through time."[2]

14.3. 27. The time of existence for the fish was about 60 million years [s.14.3. 17], which is almost equal to the whole Jurassic period. "Jurassic Period is second of three periods of the Mesozoic Era, extending from 199.6 million to 145.5 million years ago." Exact calculation gives us 54 million years. How can that species survive for such a long time? Obviously, nobody has the answer to that question. As a result they use the easiest way to explain it and decided that the fish survived – without evolutionary changes – for 60 million years.

14.3. 28. That was acceptable if someone had earlier evidence about such a species' presence in the area before 1938 [s.14.3. 17]. That can be used as proof for a survived species at one particular region. But there is no such evidence. This creates the following question. What happened in 1938? Nobody has an answer for that question, but Z-Theory explains that quite easily.

14.3. 29. As one might remember, there is the Window Effect described in detail at [12.14] and especially at [12.14. 17]. Also, figure [12.14. 1.A] gives an answer to the previous question. No evidence before 1938 was possible because of the species' absence before that year. In other words no observation of that fish type was possible before the Observation Window's first point. That window opened a bit earlier than the time of first observation (1938). By opened window I mean the first point of time when the first fish used a Z-Gate. An example for such a case is Z-Trajectory Z_3 and time point T_5 for Observation Window F [s.12.14. 1.A].

14.3. 30. More than that, there is a controversy at [14.3. 17] for the phrases "It was later discovered that these fishes were well known to the islanders, who considered the flesh

---

[1] "Seychelles." Encyclopaedia Britannica. Encyclopaedia Britannica 2008 Deluxe Edition. Chicago: Encyclopaedia Britannica, 2008.
[2] "Evolution." Encyclopaedia Britannica. Encyclopaedia Britannica 2008 Deluxe Edition. Chicago: Encyclopaedia Britannica, 2008.

edible when dried and salted" and "Rewards were offered for more specimens, and in 1952 a second (named Malania anjouanae but probably not separable from Latimeria) was obtained from near the Comoros Islands". Is it likely that one had to be hunting *for 14 years* – with "offered rewards" for "well-known" fish that "were well-known to the islanders" – before another specimen was found? That is impossible from my point of view because no islander would be able to survive on any island with such a poor hunting skill.

14.3. 31. From my point of view, there are at least two successfully completed Z-Processes (from IB point of view) at about 1938 and 1952 years. Those processes brought the fish (coelacanths) through a Z-Gate to a particular area (Z-Sector (H)), where they were discovered as a live species. Hence, there were at least two appearances of In-Gap leading to fish in our word (CE-Space). Moreover, there is nothing to explain the unsuccessful hunting between those time points.

14.3. 32. Details given above help us to determine the corresponding Z-Sector parameters and the features of the S-Gaps that exist there. Coelacanths set a time definition between 245 and 144 million years ago [s.14.3. 16]. The area of Out-Gap must be located deep below the water surface because the coelacanths "are deep-sea fishes". Hence, their natural environment was a marine environment in a deep-sea area. That is the description of corresponding Z-Sector parameters.

14.3. 33. S-Gaps present in the deep water leads to an absence of any evidence for strange events at the area of Z-Sector (H), in the air and ocean surface regions. As a result that area seems a secure one for airplanes and surface ships. But the same consideration means there is some danger to find an S-Gap below the water surface, especially for submerged craft like submarines or bathyscaphes. In the case of back-transposition, a submarine appears at a corresponding Z-Sector at the same depth and approximately 200 million years ago, vanishing from our world "without a trace".

14.3. 34. The following figure shows the correlation between probabilities of S-Gap being found at Z-Sector (H).

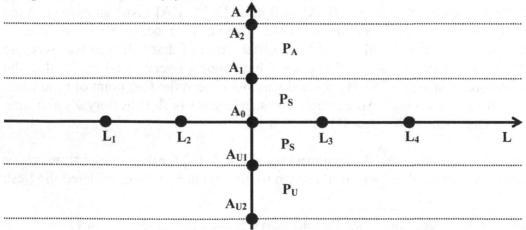

Fig. A

162

In the figure there are two axes, L and A. L-axis represents locations, and A-axis represents altitude above and below the ocean surface at Z-Sector (H). Zero altitude coincides with water surface (A_0).

14.3. 35. According to previous explanations, there are four different levels at Z-Sector (H). Two levels lie above the ocean surface and two one lie below it. Those are areas A_0-A_1, A_1-A_2, A_0-A_{U1} and A_{U1}-A_{U2}.

14.3. 36. The probability of a S-Shield presence at each level is mentioned as P_S, P_A and P_U. Levels between A_{U1} and A_1 have the same probability. They mention the case of surface ships that exist partly in air and under water. Here and later I refer to those levels as Air Layer (AL) between A_1 and A_2; Surface Layer (SL) between A_{U1} and A_1; and Underwater Layer (UL) between A_{U2} and A_{U1}.

14.3. 37. The correlation between P_A, P_S, and P_U is shown below.

$$P_A = 0 \qquad\qquad\qquad (a)$$
$$P_S = 0 \qquad\qquad\qquad (b)$$
$$P_U > 0 \qquad\qquad\qquad (c)$$

Those equations coincide with observable facts from Z-Sector (H). In other words the Dragon Effect [s.13.2. 37] happens there.

14.3. 38. **Z-Sector (C)**. One more unusual area is Z-Sector (C). Part of that Z-Sector coincides with the Himalayas as much as the Bermuda Triangle coincides with part of Z-Sector (B). That area is well seen on the map of India's north-east, going from north-west to south-east.

14.3. 39. "Himalayas. Nepali Himalaya great mountain system of Asia forming a barrier between the Tibetan Plateau to the north and the alluvial plains of the Indian subcontinent to the south. The Himalayas include the highest mountains in the world, with more than 110 peaks rising to elevations of 24,000 feet (7,300 metres) or more above sea level. One of these peaks is Mount Everest (Tibetan: Chomolungma; Chinese [Wade-Giles romanization]: Chu-mu-lang-ma Feng; Nepali: Sagarmatha), the world's highest, which reaches a height of 29,035 feet (8,850 metres). The great heights of the mountains rise above the line of perpetual snow."[1]

14.3. 40. Is there any evidence of the Dragon Effect? Certainly there is! Those are the tales and evidence about the yeti. "Abominable Snowman is Tibetan Yeti mythical monster supposed to inhabit the Himalayas at about the level of the snow line. Though reports of actual sightings of such a creature are rare, certain mysterious markings in the snow have traditionally been attributed to it. Those not caused by lumps of snow or

---

[1] "Himalayas." Encyclopaedia Britannica. <u>Encyclopaedia Britannica 2008 Deluxe Edition</u>. Chicago: Encyclopaedia Britannica, 2008.

stones falling from higher regions and bouncing across the lower slopes have probably been produced by bears. At certain gaits bears place the hindfoot partly over the imprint of the forefoot, thus making a very large imprint that looks deceptively like an enormous human footprint positioned in the opposite direction."[1]

14.3. 41. Here I give one more detail according to the definition of a snow line. "Snow line is the lower topographic limit of permanent snow cover. The snow line is an irregular line located along the ground surface where the accumulation of snowfall equals ablation (melting and evaporation). This line varies greatly in altitude and depends on several influences. On windward slopes and those facing the afternoon sun, the snow line may be as much as a kilometre (more than half a mile) higher than on opposite slopes. Over larger areas, summer temperatures and the amount of snowfall determine the position of the snow line. Where temperatures are low, as near the poles, the snow line is quite low in elevation; where temperatures are high, as near the Equator, the snow line is very high. Average altitudes of the snow line taken over large areas can be used to derive a climatic snow line, which rises or falls in altitude in response to worldwide climatic change. During glacial periods, the climatic snow line was from 600 to 1,200 m (2,000 to 4,000 feet) lower than at present."[2]

14.3. 42. Hence, the reference for an abominable snowman inhabiting the area given at [14.3. 40] as "at about the level of the snow line" gives quite a wide area with a significant altitude variation. In other words it is not just a thin line going around the mountains at a given altitude. That gives us some space for the appearance of an abominable snowman.

14.3. 43. Some readers might ask, "Why does Dragon Effect lead to such evidence only at part of Z-Sector (C)?" Generally the same evidence can be found at any point of the same Z-Sector because it has equal geographical and gravitational parameters. Moreover, as we can see in the map, Z-Sector (C) is *the only place on the Earth's surface with a combination of a high mountainous region and an NDA (from -50 to -60 mGal).*

14.3. 44. Therefore, the corresponding area for Z-Sector (C) must:

a. Be an NDA with an anomaly equal to Z-Sector (C) (from -50 to -60 mGal)
b. Be a high, mountainous area
c. Be inhabited with a particular species (abominable snowman) either in the past or in the future

14.3. 45. Evidence about stones falling "from nowhere" coincides as well with features of the Z-Gate described in detail in [14.1. 10]. As a result we have two types of independent evidence. These are observations of abominable snowman and stones

---

[1] "Abominable Snowman." Encyclopaedia Britannica. Encyclopaedia Britannica 2008 Deluxe Edition. Chicago: Encyclopaedia Britannica, 2008.
[2] "Snow line." Encyclopaedia Britannica. Encyclopaedia Britannica 2008 Deluxe Edition. Chicago: Encyclopaedia Britannica, 2008.

falling. In the case of high mountains, a lot of IB are unable to locate an exact point of a falling object because the difference of the ground level prevents them from such direct observation. Hence, they produce the easiest explanation involving the presence of a strange creature.

Physically it must be independent Z-Processes that involve separate transpositions of rocks and creatures.

14.3. 46. The following figure shows that case of the Z-Process.

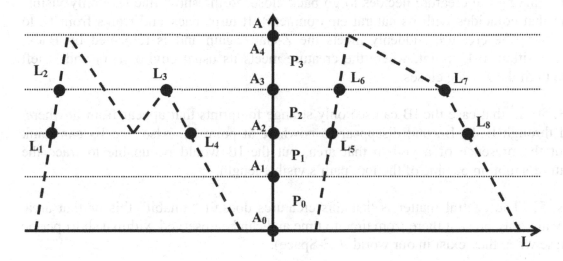

Fig. A

Figure [A] looks equal to figure [14.3. 34.A] with only one difference: it represents the positive attitudes of the mountains rising from zero altitude (A_0) to top altitude (A_4). The mountains themselves are drawn by a broken line. The left part of the figure represents the corresponding Z-Sector for Z-Sector (C). The right part represents Z-Sector (C) itself.

14.3. 47. Snow lines lie between altitudes A_2 and A_3. Hence, according to evidence from Z-Sector (C), the probability of the presence of an S-Gap between A_2 and A_3 exceeds zero. That can be shown mathematically.

$$P_0 = 0 \tag{a}$$
$$P_1 = 0 \tag{b}$$
$$P_2 > 0 \tag{c}$$
$$P_3 = 0 \tag{d}$$

In other words the area L_1-L_2-L_3-L_4 exists as a corresponding Z-Sector to Z-Sector (C), shown as the area of L_5-L_6-L_7-L_8.

14.3. 48. If such creatures inhabited the Earth's surface in the past as well as coelacanths, their fossils must be found in the high mountains at an altitude that

165

coincides with the altitude of modern evidence of yetis. The age estimation of such fossils can give us information about the time location of a corresponding Z-Sector for part of Z-Sector (C) located in the Himalayas.

14.3. 49. Furthermore suppose the yeti likes to go down the mountain. Then as it reaches point L_2, it successfully passes a Z-Gate and appears at point L_6. Walking further it is very surprised by its unusual surroundings. Reaching point L_5, it turns only to be nonplussed by absolutely unusual environment. The only thing that is familiar to the creature is the snow line, which is positioned a bit above its current location. With little thinking the creature decides to go back closer to the snow line (the only visible thing that coincides with its natural environment). It turns back and walks from L_5 to L_7, where the creature suddenly meets the Z-Gate again and is relocated by Back-Transposition to L_3 point, where the creature meets its usual environment and is left there until the Z-Gate closes.

14.3. 50. In that case the IB can see only strange footprints that appear from nowhere, lead through L_6-L_5-L_7, and disappear to nowhere at point L_7. Also, the IB can think about the presence of a yeti in that area, but the IB would be unable to trace the creature's motion outside of the foot path's visible length.

14.3. 51. The central matter is that this creatures doesn't "inhabit" this or that area. They are only present there from time to time and can be observed within a short period of time when they exist in our world (CE-Space).

14.3. 52. Nevertheless the first and last points of the visible foot path of the strange creature can be used to find the exact points where S-Gaps were located. Moreover, the presence of foot path leading "from nowhere to nowhere" ensures the possibility of a Back-Transposition by Z-Gate.

14.4 Z-Sectors and Positive Deviation Areas

14.4. 1. Previous topics discussed the phenomena of areas with Negative Deviation of S_g. It's time to turn our attention to areas with Positive Deviation. Here and later I refer to such areas as PDA (Positive Deviation Area). Those areas are marked on the map in appendix C by a red color. The more deviation that exists at a point, the darker the color that marks the point. As we can see there are a few large PDAs with the highest value of deviation. Those are following areas:

Australia and Oceania

   i.   A Large area the includes most of Oceania

South America

   j.   Western part of the continent (mountainous area)

North Atlantic

    k. A large area that includes the northern part of the ocean, part of Greenland, and the UK

Eurasia

    l. The area begins between the Black and Mediterranean seas (a mountainous area) and extends in a north-east direction close to the eastern part of the Arabian Peninsula

14.4. 2. Is there any evidence from those areas about strange events? Certainly there is, with the most famous among them the Z-Sector (J). It is easy to guess which part of that area has certain evidence. Of cause I mean the Nazca Lines.

14.4. 3. "Nazca Lines are groups of large line drawings and figures that appear, from a distance, to be etched into the earth's surface on the arid Pampa Colorada ('Coloured Plain,' or 'Red Plain'), northwest of the city of Nazca in southern Peru. They extend over an area of nearly 190 square miles (500 square km).

14.4. 4. "Constructed more than 2,000 years ago by the people of the Nazca culture (c. 200 BC–AD 600), the Nazca Lines include about 70 images of plants and animals – such as a monkey (some 360 feet [110 metres] long), a killer whale (210 feet [65 metres]), a bird resembling a condor (443 feet [135 metres]), a hummingbird (165 feet [50 metres]), a pelican (935 feet [285 metres]), a spider (150 feet [46 metres]), and various flowers, trees, and other plants – as well as geometric shapes, including triangles, trapezoids, and spirals. Virtually indecipherable from ground level, they are plainly visible from the air.

14.4. 5. "Since their discovery in the 1920s, the lines have been variously interpreted, but their significance remains largely shrouded in mystery. The American historian Paul Kosok observed the lines from an airplane in 1941 and hypothesized that they were drawn for astronomical purposes. Maria Reiche, a German translator who spent years studying the site and lobbying for its preservation, also concluded that it was a huge astronomical calendar and that some of its animal sketches were modeled after groupings of stars in the night sky. In 1967, however, the American astrophysicist Gerald Hawkins found no correlation between changes in the celestial bodies and the design of the Nazca Lines. Some archaeologists believe that the site is either a cluster of sacred paths or an outdoor temple.

14.4. 6. "The Nazca Lines are preserved naturally by the region's dry climate and by winds that sweep sand out of their grooves. They are not protected from the passage of motor vehicles over them, however, and tire tracks and roadways have marred the otherwise pristine plain. UNESCO added the Nazca site to its World Heritage List in 1994."[1]

14.4. 7. I put here one more quotation that describes the country itself. "Peru is officially Republic of Peru, Spanish Republica del Peru country in South America. Except for the Lake Titicaca basin in the southeast, its borders lie in sparsely populated zones. The boundaries with Ecuador to the northwest, Bolivia to the southeast, and Chile to the south run across the high Andes, whereas the borders with Colombia to the northeast and Brazil to the east traverse lower ranges or tropical forests. Peru's land area is supplemented by territorial waters, reaching 200 miles (320 kilometres) into the Pacific, on the west, that are claimed by Peru."[2]

14.4. 8. There is additional evidence from South America. Mr. Berlitz mentions "strange" little gold figures that were found in different locations of the continent (including Republic of Columbia) in his famous work "The Bermuda triangle". Generally figures show long body with delta wings and tail. Most researches believe these figures represent birds. But there is no bird with vertical tail. Mr. Berlitz gives some proves for different point of view the creators of the figures tried to reconstruct shape of modern super-sonic aircrafts.

14.4. 9. Hence, we have the following evidence ("coincidences") from Z-Sector (J):

    a. A strong PDA (more than +60 mGal)
    b. High mountainous area (Andes)
    c. Strange lines visible only from flying aircraft
    d. Strange, birdlike figures with vertical tails

14.4. 10. The following figure shows characteristics of that area according to Z-Gate possibilities.

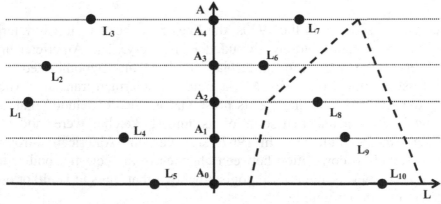

Fig. A

---

[1] "Nazca Lines." Encyclopaedia Britannica. Encyclopaedia Britannica 2008 Deluxe Edition. Chicago: Encyclopaedia Britannica, 2008.
[2] "Peru." Encyclopaedia Britannica. Encyclopaedia Britannica 2008 Deluxe Edition. Chicago: Encyclopaedia Britannica, 2008.

The figure has same the meaning as [14.3. 46.A] regarding axes and points. Negative values of the L axis show locations before Z-Process (left half of the figure), and positive ones show locations after Z-Process (right half of the figure). The broken line at positive locations represents mountains.

14.4. 11. As shown before [14.2. 60; 14.2. 61], a corresponding Z-Sector for Z-Sector (J) must have the same value of a gravity anomaly. As a result different S-Outlines cross both areas at different altitudes (from A_0 to A_4 according to the figure). A_0 means zero level of altitude (ocean surface level).

14.4. 12. As we can see in the picture, there are some possible points that can be used by Z-Process. Those are points that have corresponding S-Gaps located only *above* the surface of the mountains. Those are points L_6 and L_7. They have corresponding points for S-Gaps at L_2 and L_3 accordingly.

14.4. 13. Other points have no chance to be used for a Z-Gate because each time any system tries to use them, that system must be crushed by solid rock inside the mountains at any attempt to use In-Gate. That happens because the moving system keeps its velocity equal before and after Z-Gate (a law of conservation restriction). Those corresponding points are shown in the figure as L_1-L_8, L_4-L_9, and L_5-L_{10}. All of them are Z-Gates that cannot be passed successfully, especially by any system using high velocity.

14.4. 14. The only possible way to use a Z-Gate successfully is that the moving system must use only Z-Gates that have both corresponding points above Earth's surface. According to that law only L_2-L_6 and L_3-L_7 transpositions are possible as successful ones.

14.4. 15. Which type of system (object) can use such Z-Gates? Obviously, that can be only a system (object) that is able to reach a high enough altitude. As a result we have only one possible system that meets that condition: flying things such as an airplane.

14.4. 16. Hence, the corresponding Z-Sector must be a PDA with the same value of deviation and the presence of flying artificial objects. Therefore, any aircraft that disappeared without a trace above the same PDA (it can be Z-Sector (J) itself, too) and successfully passed through a Z-Gate appears above a strange endless mountain. Obviously, the pilot (or crew) has no idea about their new location and tries to land. Also an IB (that can be an aborigine) sees a strange flying object (or a bird with a vertical tail) that appeared from nowhere.

14.4. 17. The activity of the crew after a more or less successful landing can be extended from attempts to locate their position to attempts to show their location for rescue squads. What can a man do to attract attention to an aircraft crew? Nothing more than drawing huge signs of his presence in any possible way. That happens because a pilot thinks that the more unusual the symbol created, the higher the probability of it being noticing from an aircraft.

14.4. 18. Moreover, what can a crash-landed crew do to help the rescue squads have a successful landing not far from their location? Obviously, they try to build a runway because they are awaiting rescue.

14.4. 19. From my point of view, that is quite enough to explain the Nazca Lines' cause of presence and the absence of any relationship between the lines and any other possible explanation of their presence [r.14.4. 5]. That also gives us additional evidence of a NTT possibility. That is the only way to explain the evidence of super-sonic aircraft presence at that time [r.14.4. 8].

14.4. 20. **Z-Sector (K)**. That Z-Sector covers a large area with the maximal gravity deviation at about +50 mGal. The most famous part of that region lies at it south-east point where the gravity anomaly reaches a value of about +20 mGal (according to the map in appendix C). That is the northern part of the UK, and the legendary Loch Ness is located there.

14.4. 21. "Ness, Loch is the lake, lying in the Highland council area, Scotland. With a depth of 788 feet (240 metres) and a length of about 23 miles (36 km), Loch Ness has the largest volume of fresh water in Great Britain. It lies in the Glen Mor – or Great Glen, which bisects the Highlands – and forms part of the system of waterways across Scotland that civil engineer Thomas Telford linked by means of the Caledonian Canal (opened 1822).

14.4. 22. "The watershed of Loch Ness covers more than 700 square miles (1,800 square km) and comprises several rivers, including the Oich and the Enrick. Its outlet is the River Ness, which flows into the Moray Firth at Inverness. Seiches (surface oscillations), caused by differential heating, are common on the loch. The sharp rise and fall of the level of the loch is one reason for the scanty flora of the waters; another reason is the great depths of the loch near the shoreline. The abyssal fauna is also sparse.

14.4. 23. "Like some other very deep lochs in Scotland and Scandinavia, Loch Ness is said to be inhabited by an aquatic monster. Many sightings of the so-called Loch Ness monster have been reported, and the possibility of its existence – perhaps in the form of a solitary survivor of the long-extinct plesiosaurs – continues to intrigue many."[1]

14.4. 24. We must pay special attention to the quotation mentioned above because each piece of information mentions a lot of details about Z-Appearances.[2] First of all there is a very strange explanation about "Seiches (surface oscillations), caused by differential heating, are common on the loch" [s.14.4. 22]. That leads to the following questions.

---

[1] "Ness, Loch." Encyclopaedia Britannica. Encyclopaedia Britannica 2008 Deluxe Edition. Chicago: Encyclopaedia Britannica, 2008.
[2] See dictionary (Appendix A).

What is that unusual cause of "differential heating"? Which process is able to heat a large lake, especially for a temperature that caused significant surface oscillations? Which process is able to cool that same volume of lake water to reach a "fall of the level of the loch" [s.14.4. 22]? More than that, the next question appears as a consequence of the previously mentioned one. What is the strange combination of cooling and heating processes in the lake that causes a "sharp rise and fall of the level of the loch" [s. 14.4. 22]?

14.4. 25. From my point of view, the thermal cause used as an explanation of surface oscillations is beneath criticism. According to Z-Theory there is a more applicable explanation of such a process. That process was described in detail at [12.12–12.13] as the Surge Effect. In the case of Loch Ness, the Z-Current explains all water phenomena including surface oscillations, heating, and cooling. All phenomena can be explained by one process (Z-Current) that occurs only in liquids.

14.4. 26. The following figure shows an imagined cross-section of the lake according to two corresponding Z-Sectors; one of them is partly occupied by the lake.

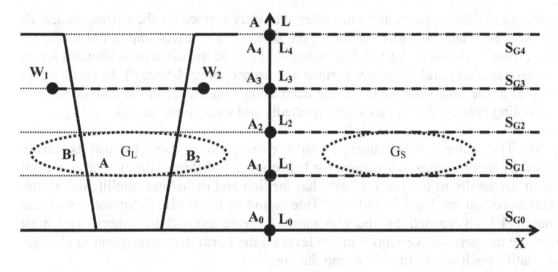

Fig. A

The figure shows two areas of corresponding Z-Sectors. Axis X represents projection of S-Outlines to some altitude that is mentioned as zero one (A_0). The Negative part of the X axis coincides with part of Z-Sector (K), which is occupied by the lake. A positive part of the same axis represents the corresponding Z-Sector for Z-Sector (K).

Different S-Outlines corresponding with different altitudes and water levels are mentioned as (S_{G0}- S_{G4}), (A_0-A_4), and (L_0-L_4) accordingly. The average water level in the lake is shown as a broken line, W_1-W_2 (which coincides with A_3, L_3, S_{G3}).

14.4. 27. Two ovals between S_{G1} and S_{G2} represent two corresponding S-Gaps [s.14.4. 26.A]. Oval G_L represents the S-Gap of the lake, and G_S represents the S-Gap of the

171

corresponding Z-Sector. Moreover, the S-Gap shown in the lake (G_L) uses more space than the lake permits at some particular depth (where the S-Gap exists). As a result it splits into two parts, A and B. Part A is located in the lake water, and part B is located in the ground surrounding the lake. Part B is separated into two zones, B_1 and B_2, because of any natural pond shape.

14.4. 28. If a Z-Gate appears at the time when the water surface on the lake and the corresponding Z-Sector has equal levels, no visible effects (water surface effect) are present at the lake. Hence, the Z-Gate is able to be present as long as possible without any evidence (from an IB located at the lake shore) of strange observation [s. 12.12].

14.4. 29. If a Z-Gate appears at a time when the water surface on the corresponding Z-Sector is raised more than the water surface of the lake, the Z-Current appears from G_S to G_L [s. 12.12]. For example, it can be the difference between levels L_3 (the lake surface) and L_4 (water surface at corresponding Z-Sector). In that case an IB can see the increasing water level in the lake coinciding with a warming process that appears "unexpectedly and with no reason" [s. 12.12]. In other words the IB sees "Seiches (surface oscillations), caused by differential heating" [s.14.4. 22].

14.4. 30. If a Z-Gate appears at a time when the water surface on the corresponding Z-Sector falls less than the water surface of the lake, the Z-Current appears at a counter direction from G_L to G_S [s. 12.12]. For example, it can be the difference between levels L_3 (the lake surface) and L_2 (water surface at corresponding Z-Sector). In that case an IB (located at the lake shore) can see the decreasing water level in the lake coinciding with a cooling process that appears "unexpectedly and with no reason" [s. 12.12].

14.4. 31. The water level changing with a corresponding Z-Sector can be easily explained by natural tides that happen (or happened at a different time) at any point of the Earth. Generally as long as the Earth has the Sun and its natural satellite the Moon, the tides occur on the Earth's surface. "Tide is any of the cyclic deformations of one astronomical body caused by the gravitational forces exerted by others. The most familiar are the periodic variations in sea level on the Earth that correspond to changes in the relative positions of the Moon and the Sun."[1]

14.4. 32. The correlation between the lake tides caused by Z-Current and the natural tides in the present time cannot be strong because the corresponding Z-Sector's possible location can be quite different from the Z-Sector (K) location (characteristics of S-Outline).

14.4. 33. As we can see at [14.4. 26.A], the location of both S-Gaps are independent from the physical water level at both the corresponding Z-Sectors and the direction of Z-Current.

---

[1] "Tide." Encyclopaedia Britannica. Encyclopaedia Britannica 2008 Deluxe Edition. Chicago: Encyclopaedia Britannica, 2008.

14.4. 34. This leads to the next serious question. What happens if any BS (creature) [s.13.1. 2] uses the same Z-Gate as is the one used by water? The answer is this: The creature disappears from its natural location and appears from nowhere at the corresponding Z-Sector. If the BS uses the same Z-Gate again, it disappears from the strange location and appears from nowhere at its natural location.

14.4. 35. According to [14.4. 23], "Many sightings of the so-called Loch Ness monster have been reported, and the possibility of its existence – perhaps in the form of a solitary survivor of the long-extinct plesiosaurs – continues to intrigue many." Hence, there are reports about the observation of a plesiosaur (Dragon Effect [s.13.2]). Does it make any sense that to be the answer? The coincidence of two effects explained above (Surge Effect and Dragon Effect) occurring in one particular lake?

14.4. 36. From my point of view, such a coincidence gives us the strongest proof for the whole of Z-Theory. Hence, the presence of a plesiosaur can be explained as well as any other Dragon Effects (e.g., the presence of coelacanths and the observation of abominable snowmen). The same explanation was mentioned at [14.3. 51].

14.4. 37. Just as abominable snowmen never "inhabit" a particular area, the Loch Ness plesiosaur never "inhabits" the lake. Something that is seen by an IB is nothing more than a number of the same observations of a type of reptile successfully that passed through a Z-Gate. Obviously, the same reptiles have the same behavior and look from like the same creature. That caused tale about a single monster.

14.4. 38. Biologically, no creature is able to survive in a local pond for about 65 million years. That is the time between the last evidence of plesiosaurs' presence to the present time [s.13.3. 21]. Moreover, "The sharp rise and fall of the level of the loch is one reason for the scanty flora of the waters; another reason is the great depths of the loch near the shoreline. The abyssal fauna is also sparse." [s.14.4. 22]. In other words there is not enough of a feeding source for a permanent presence of a heavy weighted reptile to survive.

14.4. 39. Nevertheless the creatures can survive for some time in the lake, especially when their presence time is less than the time of a Z-Gate presence. Here and later I refer to the time between first and last point of time of a Z-Gap presence according to the IB's TKD as the Z-Window. Therefore, as soon as the creature reaches the underwater Z-Gap in the lake, it disappears from the lake. Any other attempts to find the creature in the lake later have no success because of physical absence of the creature in that time.

14.4. 40. Such a sequence of observations caused a lot of people to be skeptic about any "tail" of "a monster". That happens because the creature can be observed in the following case.

$$P_O = P_G \cdot P_C \qquad \text{(a)}$$

Equation A means that the probability of observation of the Loch Ness monster (P_O) equals the multiplication of two independent probabilities. Those are the probability of Z-Gate presence in the lake (P_G) and the probability of using that gate by a creature (P_C).

14.4. 41. Equation [14.4. 40.a] gives an answer to the question about the different probability of observation of a creature and the surface oscillations. In the case of surface oscillations, that equation transforms to following one.

$$P_{OS} = P_G \cdot P_T \tag{a}$$

Equation A means that the probability of observation of Loch Ness's surface oscillations (P_{OS}) equals the product of two independent probabilities. Those are the probability (P_G) of Z-Gate presence in the lake and the probability (P_T) of the tide happening at the corresponding Z-Sector.

14.4. 42. Because the probability of the tide is much higher than the probability of using a Z-Gate by a monster, then P_T[1] is much greater than P_C. As a result P_{OS} is much greater than P_O ($P_{OS} \gg P_O$). That fully coincides with real observations [14.4. 21–14.4. 23]: the surface oscillations occur much more frequently than the observation of the monster.

14.4. 43. Looking again at figure [14.4. 26.A], we can see this is only part of S-Gap can be located in the lake (area 'A'). Hence, despite the S-Gap size it is penetrable by a creature only in area A because no creature is able to penetrate solid rock or the ground of the Earth outside of the lake. That possibility leads to the concentration of creatures successfully passing the Z-Gate in the lake instead of the surrounding areas.

14.4. 44. Moreover, an S-Gap in a shape like the one shown at figure [14.4. 26.A] explains the easiness of a creature's disappearance. As we can see the creature needs only to dive from the water surface (level L_3, A_3) to some depth of the nearest S-Gap point (level L_2, A_2). In other words a creature does not need to find a difficult way that leads back to the S-Gap.

14.4. 45. According to [14.4. 23], "Like some other very deep lochs in Scotland and Scandinavia, Loch Ness is said to be inhabited by an aquatic monster." It's time to see the map (appendix C) again. The map shows the same value of Z-Sector (K) gravity deviation (about +20 mGal) for the entire area of Scotland and West Scandinavia (especially the south-west and north-west parts of Scandinavia). That is the edge of the same Z-Sector. Therefore, the type of observable events from the same Z-Sector looks equal. From my point of view that gives an additional proof for Z-Theory.

14.4. 46. Hence, putting all things together in reference to Z-Sector (K), we have the following aspects for the corresponding Z-Sector.

---

[1] That difference occurs because the tide is a continuous process that happens on the Earth's surface.

a. It lies at an area with a gravity anomaly of about +20 mGal
b. The area is covered by water where significant tides occur at a regular basis (possibly a seashore or a ocean shore)
c. The area is inhabited by aquatic creatures using the water (plesiosaurs)
d. The area is distanced from the present time, approximately from 215 million to 80 million years ago (based on plesiosaurs' fossil information [s.13.3. 21])

14.4. 47. One might come to have the following question. How it is possible for the Z-Process to find locations of the Earth (and its corresponding Z-Sectors) distanced a million years from each other? The following chapter gives detailed answers for that question and explains how they are related to those characteristics of the Z-Process. As a result it's best to turn our attention away from the beloved planet and trace the possible path of Z-Processes in deep space.

Chapter 15. Galaxy-related Appearances

15.1 Relation with the Sun's Orbit

15.1. 1. Previous chapters of the work explain Z-Processes and their consequences according to the Earth's nearest celestial bodies (the Sun and the Moon) and the Earth's gravity anomalies. All those explanations were given relative to S_g value and its deviation at different space points around the Earth.

15.1. 2. Besides those bodies producing a gravity field, there are a lot of massive objects in the galaxy, and each of them also produces a field. Most of those objects are others stars, forming a huge and very complex structure that we call galaxy. Each object in the galaxy produces it own gravity field and interacts with any other object by gravitation.

15.1. 3. "Milky Way Galaxy is a large spiral system consisting of several billion stars, one of which is the Sun. It takes its name from the Milky Way, the irregular luminous band of stars and gas clouds that stretches across the sky. Although the Earth lies well within the Galaxy, astronomers do not have as clear an understanding of its nature as they do of some external star systems. A thick layer of interstellar dust obscures much of the Galaxy from scrutiny by optical telescopes, and astronomers can determine its large-scale structure only with the aid of radio and infrared telescopes, which can detect the forms of radiation that penetrate the obscuring matter."[1]

15.1. 4. The size of the galaxy and the location of the Sun in modern astronomy has the following information. "The most distant stars and gas clouds of the system that have had their distance determined reliably lie roughly 72,000 light-years from the galactic centre, while the distance of the Sun from the centre has been found to be approximately 27,000 light-years."[2]

15.1. 5. Just as the planets of the solar system move around the center (the Sun), all stars in galaxy move around the galaxy core (the center of the galaxy). The Sun has its own velocity as a relative motion between the star and the galaxy core. Modern measurements give us following result.

15.1. 6. "Since the direction of the centre of the Galaxy is well established by radio measurements and since the galactic plane is clearly established by both radio and optical studies, it is possible to determine the motion of the Sun with respect to a fixed frame of reference centred at the Galaxy and not rotating (i.e., tied to the external galaxies). The value for this motion is generally accepted to be 225 km/sec in the

---

[1] "Milky Way Galaxy." Encyclopaedia Britannica. Encyclopaedia Britannica 2008 Deluxe Edition. Chicago: Encyclopaedia Britannica, 2008.
[2] "Milky Way Galaxy." Encyclopaedia Britannica. Encyclopaedia Britannica 2008 Deluxe Edition. Chicago: Encyclopaedia Britannica, 2008.

direction $L^{II} = 90°$. It is not a firmly established number, but it is used by convention in most studies."[1]

15.1. 7. Additionally, I present here two more notions used in modern astronomy, mostly for dear readers who are not familiar with that area. These are the astronomical length units light-year and parsec.

15.1. 8. "Light-year in astronomy, the distance traveled by light moving in a vacuum in the course of one year, at its accepted velocity of 299,792,458 metres per second (186,282 miles per second). A light-year equals about 9.46053×10^{12} km (5.878×10^{12} miles), or 63,240 astronomical units. About 3.262 light-years equal one parsec"[2]

15.1. 9. "Parsec is unit for expressing distances to stars and galaxies, used by professional astronomers. It represents the distance at which the radius of the Earth's orbit subtends an angle of one second of arc; thus a star at a distance of one parsec would have a parallax of one second, and the distance of an object in parsecs is the reciprocal of its parallax in seconds of arc. For example, the nearest triple-star system, Alpha Centauri, has a parallax of 0.753 second of arc; hence, its distance from the Sun and the Earth is 1.33 parsec. One parsec equals 3.26 light-years, which is equivalent to 3.09×10^{13} km (1.92×10^{13} miles). In the Milky Way Galaxy, wherein the Earth is located, distances to remote stars are measured in terms of kiloparsecs (1 kiloparsec = 1,000 parsecs). The Sun is at a distance of 8.5 kiloparsecs from the centre of the Milky Way system. When dealing with other galaxies or clusters of galaxies, the convenient unit is the megaparsec (1 megaparsec = 1,000,000 parsecs). The distance to the Andromeda Galaxy (Messier 31) is about 0.7 megaparsec. Some galaxies and quasars have likely distances on the order of about 3,000 megaparsecs, or 9,000,000,000 to 10,000,000,000 light-years."[3]

15.1. 10. As we can see, the size of the galaxy and the surrounding space is so tremendous that it needs to be scaled only in terms of the appropriate measure units. Usual length units used for measuring the size of the Earth and the distance between points on its surface are absolutely useless on the scale of the universe. Nevertheless despite the universe's size any physical principles keep their power at any point of space, helping us to understand distant events as well as near ones.

15.1. 11. According to the Sun's velocity around the galaxy's center and its distance from the galaxy center, it's quite possible to calculate the period of time that the star takes for one full revolution. Here and later I refer to that period of time as a Sun Galaxy Year (SGY) (Star Galaxy Year in general condition).

[1] "Milky Way Galaxy." Encyclopaedia Britannica. Encyclopaedia Britannica 2008 Deluxe Edition. Chicago: Encyclopaedia Britannica, 2008.
[2] "Light-year." Encyclopaedia Britannica. Encyclopaedia Britannica 2008 Deluxe Edition. Chicago: Encyclopaedia Britannica, 2008.
[3] "Parsec." Encyclopaedia Britannica. Encyclopaedia Britannica 2008 Deluxe Edition. Chicago: Encyclopaedia Britannica, 2008.

15.1. 12. **Sun Galaxy Year calculation**. To simplify the following calculation, I use the notion of Star Light Velocity Rate (SLVR). It shows how many times the speed of the star is less than the speed of light. For the Sun moving around the galaxy center, we have the following value [r.15.1. 6; 15.1. 8].

$$299{,}792{,}458 \ [m/s] \ / \ 225{,}000 \ [m/s] = 1{,}332.4 \qquad\qquad (a)$$

Hence, the Sun moving with its current speed covers a distance equal to 1 light-year in 1,332.4 years (because its speed is 1,332.4 times less than the speed of light).

The full length of the Sun's galaxy orbit can be calculated as the length of a circle with radius equal to the distance between the Sun and the galaxy center [s.15.1. 4].

$$L = 2 \cdot \pi \cdot r = 2 \times 3.14 \times 27{,}000 \ [\text{light-years}] = 169{,}560 \ [\text{light-years}] \qquad (b)$$

Therefore, the time of SGY can be easily calculated by multiplying the length of the Sun's orbit in SLVR.

$$SGY = L \times SLVR = 169{,}560 \ [\text{light-years}] \times 1{,}332.4 = 225{,}921{,}744 \ [\text{years}] \qquad (c)$$

For further references I use SGY as 226 million years (2.26×10^8 years).

15.1. 13. Therefore, every 226 million years the Sun takes the same position relative to the galaxy center. If the orbit of the Sun is not an exact circle and forms an ellipse close to a right circle's distance from the galaxy center, it varies slowly on a scale of millions of years. In that case the S_g produced by the galaxy at each point of the Sun's location has the same variation in time and adds various values to the full S_g calculated at each star's location (galaxy influences). Those values count again and keep equally within each full circle of the Sun's motion (SGY) in each location.

Figure [10.2. 1.A] can be used as an example for such a motion, keeping in mind that this elliptical trajectory of the Sun is very close to right circle. The ellipse of [10.2. 1.A] is highly stretched horizontally for explanation purposes only.

In cases of the Sun's motion, figure [10.2. 1.A] represents the galaxy center as P and the Sun's orbit as A-B-C-D-A trajectory. Distances PB and PD are equal to each other as already mentioned (see the explanation of the figure).

15.1. 14. Suppose the Sun occupies location B at present. In that case the equation [5.1. 10.a] has an exact equality once during every 226-million-year period. In other words each location of the Sun and its neighboring celestial bodies (planets, etc.) has the same value of S_g produced by the galaxy with a period of time equal to SGY (226 million years).

15.1. 15. Hence, the SGY appears as a main period of time that separates corresponding S-Gaps on the galaxy scale. Moreover, at each moment of time the previous orbit of the

Sun around the galaxy's center can be used as a corresponding point of time for S-Gaps. That happens for any number of SGY [s.10.2. 12.b]. What does it mean for S-Gaps according to the Earth? That is, what happened on the Earth 226 minions years ago? A lot of information about that time gives us fossils of Triassic Period. That information is well used to create a description of the period.

15.1. 16. "Triassic Period is in geologic time, the first period of the Mesozoic Era. It began 251 million years ago, at the close of the Permian Period, and ended 199.6 million years ago, when it was succeeded by the Jurassic Period."

15.1. 17. The Triassic Period marked the beginning of major changes that were to take place throughout the Mesozoic Era, particularly in the distribution of continents, the evolution of life, and the geographic distribution of living things. At the beginning of the Triassic, virtually all the major landmasses of the world were collected into the supercontinent of Pangea. Terrestrial climates were predominately warm and dry (though seasonal monsoons occurred over large areas), and the Earth's crust was relatively quiescent. At the end of the Triassic, however, plate tectonic activity picked up, and a period of continental rifting began. On the margins of the continents, shallow seas, which had dwindled in area at the end of the Permian, became more extensive; as sea levels gradually rose, the waters of continental shelves were colonized for the first time by large marine reptiles and reef-building corals of modern aspect."[1]

15.1. 18. "At the beginning of the Triassic Period, the present continents of the world were grouped together into one large C-shaped supercontinent named Pangea (see the map). Covering about one-quarter of the Earth's surface, Pangea stretched from 85° N to 90° S in a narrow belt of about 60° of longitude. It consisted of a group of northern continents collectively referred to as Laurasia and a group of southern continents collectively referred to as Gondwana. The rest of the globe was covered by Panthalassa, an enormous world ocean that stretched from pole to pole and extended to about twice the width of the present-day Pacific Ocean at the Equator. Scattered across Panthalassa within 30° of the Triassic Equator were islands, seamounts, and volcanic archipelagoes, some associated with deposits of reef carbonates now found in western North America and other locations."[2]

15.1. 19. "Worldwide climatic conditions during the Triassic seem to have been much more homogeneous than at present. No polar ice existed. Temperature differences between the Equator and the poles would have been less extreme than they are today, which would have resulted in less diversity in biological habitats."[3]

---

[1] "Triassic Period." Encyclopaedia Britannica. Encyclopaedia Britannica 2008 Deluxe Edition. Chicago: Encyclopaedia Britannica, 2008.
[2] "Triassic Period." Encyclopaedia Britannica. Encyclopaedia Britannica 2008 Deluxe Edition. Chicago: Encyclopaedia Britannica, 2008.
[3] "Triassic Period." Encyclopaedia Britannica. Encyclopaedia Britannica 2008 Deluxe Edition. Chicago: Encyclopaedia Britannica, 2008.

15.1. 20. "Fossils of marine reptiles such as the shell-crushing placodonts (which superficially resembled turtles) and the fish-eating nothosaurs occur in the Muschelkalk, a rock formation of Triassic marine sediments in central Germany. The nothosaurs, members of the sauropterygian order, did not survive the Triassic, but they were ancestors of the large predatory plesiosaurs of the Jurassic. The largest inhabitants of Triassic seas were the early ichthyosaurs, superficially like dolphins in profile and streamlined for rapid swimming. These efficient hunters, which were equipped with powerful fins, paddle-like limbs, a long-toothed jaw, and large eyes, may have preyed upon some of the early squidlike cephalopods known as belemnites. There also is evidence that these unusual reptiles gave birth to live young."[1]

15.1. 21. "Flying reptiles. Some of the earliest lizards may have been the first vertebrates to take to the air. Gliding lizards, such as the small Late Triassic Icarosaurus, are thought to have developed an airfoil from skin stretched between extended ribs, which would have allowed short glides similar to those made by present-day flying squirrels. Similarly, Longisquama had long scales that could have been employed as primitive wings, while the Late Triassic Sharovipteryx was an active flyer and may have been the first true pterosaur (flying reptile). All these forms became extinct at the end of the Triassic, their role as fliers being taken over by the later pterosaurs of the Jurassic and Cretaceous."[2]

15.1. 22. As one might remember, pterosaurs inhabited the Earth from 248 million to 65 million years ago [r.13.2. 36]. The beginning of that period matches the SGY (226 million years). Coelacanths also had the same time of existence, from 245 to 144 million years ago [s.14.3. 16]. According to the phenomena mentioned in the previous chapter of an aquatic monster supposedly inhabiting Loch Ness, it coincides with "The nothosaurs, members of the sauropterygian order, did not survive the Triassic, but they were ancestors of the large predatory plesiosaurs of the Jurassic" [s.15.1. 20].

15.1. 23. As we can see nearly all types of strange creatures that appear at any modern Z-Sector coincide with the same types of monsters from Earth's history 226 million years ago. All of them inhabited the Earth at that time.

15.2 Previous Circle

15.2. 1. Equation [10.2. 11.b] gives the relation between time points of S-Gaps and their possible locations in time. Applying that equation for SGY, we have a result of 226 million years ago for n=-1 [r.15.1. 12]. That happens because the period of SGY was calculated at [15.1. 12] as an exact value. Using that result to calculate n, its value becomes equal to minus one.

---

[1] "Triassic Period." Encyclopaedia Britannica. Encyclopaedia Britannica 2008 Deluxe Edition. Chicago: Encyclopaedia Britannica, 2008.
[2] "Triassic Period." Encyclopaedia Britannica. Encyclopaedia Britannica 2008 Deluxe Edition. Chicago: Encyclopaedia Britannica, 2008.

15.2. 2. In the case of one more SGY and a n=-2 equation [10.2. 11.b], that gives us the following result: $t_2 = t_1 - 2 \cdot 226,000,000 = t_1 - 452,000,000$. In cases of t_1 coinciding with the present time and having a zero value (the beginning of time axis), t_2 becomes equal to -452 million years (or 452 million years ago). It's quite possible to calculate any period of time for corresponding S-Gaps using the same method (for n equal to -3, -4, -5, etc.).

15.2. 3. What was on the Earth 452 million years ago? The answer is in the fossils of that geologic period. According to geologic time the Ordovician Period occurred at that time. "Ordovician Period in geologic time, the second period of the Paleozoic Era. It began 488.3 million years ago, following the Cambrian Period, and ended 443.7 million years ago, when the Silurian Period began. Ordovician rocks have the distinction of occurring at the highest elevation on Earth – the top of Mount Everest."[1]

15.2. 4. "The Ordovician Period ushered in significant changes in plate tectonics, climate, and biological systems. Rapid seafloor spreading at oceanic ridges fostered some of the highest global sea levels in the Phanerozoic Eon. As a result, continents were flooded to an unprecedented level, with North America almost entirely underwater at times. These seas deposited widespread blankets of sediment that preserved the extraordinarily abundant fossil remains of marine animals. Numerical models of the Ordovician atmosphere estimate that levels of carbon dioxide were several times higher than today. This would have created warm climates from the Equator to the poles; however, extensive glaciation did occur for a brief time over much of the Southern Hemisphere at the end of the period."[2]

15.2. 5. "Until the discovery in 1999 of Early Cambrian vertebrates in South China, the oldest generally accepted vertebrates were known from the Ordovician. The first examples were two genera of primitive fishes, described by American geologist Charles Doolittle Walcott in the late 19th century, from the Upper Ordovician Harding Sandstone of Colorado."[3]

15.2. 6. "All of these fossils are interpreted to be agnathans, or jawless fishes. The environment in which they lived continues to be disputed, although the interpreted environment at all of the localities is similar. At all three locations, sediments were laid down in very shallow marine to marginal marine environments, possibly with low salinity as found in lagoons and estuaries. The fishes are interpreted to have fed on organic matter on the seafloor."[4]

---

[1] "Ordovician Period." Encyclopaedia Britannica. Encyclopaedia Britannica 2008 Deluxe Edition. Chicago: Encyclopaedia Britannica, 2008.
[2] "Ordovician Period." Encyclopaedia Britannica. Encyclopaedia Britannica 2008 Deluxe Edition. Chicago: Encyclopaedia Britannica, 2008.
[3] "Ordovician Period." Encyclopaedia Britannica. Encyclopaedia Britannica 2008 Deluxe Edition. Chicago: Encyclopaedia Britannica, 2008.
[4] "Ordovician Period." Encyclopaedia Britannica. Encyclopaedia Britannica 2008 Deluxe Edition. Chicago: Encyclopaedia Britannica, 2008.

15.2. 7. "Although no fossils of land animals are known from the Ordovician, burrows from the Late Ordovician of Pennsylvania have been interpreted as produced by animals similar to millipedes. A millipede-like organism is inferred because the burrows occur in discrete size classes, are bilaterally symmetrical, and were backfilled by the burrowing organism. The burrows are found in a preserved soil and are associated with carbonate concretions that precipitated within the soil, indicating that the burrows were produced at the time of soil formation. The presence of plants and possibly arthropods suggests that Ordovician terrestrial ecosystems may have been more extensive and complex than generally thought."[1]

15.2. 8. The most useful evidence (according to the aim of this work) is "the extraordinarily *abundant* fossil remains of marine animals" [s.15.2. 4], "jawless fishes" were "interpreted to have fed on organic matter on the seafloor" [s.15.2. 6], and "no fossils of land animals are known from the Ordovician" [s.15.2. 7].

15.2. 9. Hence, the Earth at that time was inhabited mostly by marine life and had almost no land animals. In other words the probability of using S-Gaps by marine life forms was much higher than land-based life forms. The second probability is almost equal to zero. Therefore, we have the chance to find some traces of that period's species only in the water. What's more, that must not be "strange looking fishes" or "degraded specimen of fish" caused by human pollutions to the ocean. It must be only water specimens that coincide with fossils of the Ordovician Period, just as the coelacanth body structure coincides with fossils of the same species.

15.2. 10. Generally a modern IB can see only a water Dragon Effect from the second SGY because of the absence of flying reptiles or land animals at the Ordovician Period.

15.3 The Elder Circles

15.3. 1. Going further in the past, we can calculate the time of the third SGY circle. An exact calculation gives the value of 678 million years ago (equal to SGY multiplied by three). According to geologic time that point belongs to the Late Proterozoic Era. "Proterozoic Eon is the younger of the two divisions of Precambrian time, extending from 2.5 billion to 540 million years ago. It is often divided into the Early Proterozoic Era (2.5 to 1.6 billion years ago), the Middle Proterozoic Era (1.6 billion to 900 million years ago), and the Late Proterozoic Era (900 to 540 million years ago). Proterozoic rocks have been identified on all the continents and often constitute important sources of metallic ores, notably of iron, gold, copper, uranium, and nickel. It is thought that the many small protocontinents that had formed during early Precambrian time coalesced into one or several large landmasses by the initial segment of the Proterozoic. Rocks of

---

[1] "Ordovician Period." Encyclopaedia Britannica. Encyclopaedia Britannica 2008 Deluxe Edition. Chicago: Encyclopaedia Britannica, 2008.

the Proterozoic contain many definite traces of primitive life-forms – e.g., the fossil remains of bacteria and blue-green algae."[1]

15.3. 2. The Precambrian time mentioned above has the following characteristics. "Precambrian time is period of time that extends from a little more than 3.9 billion years ago, which is the approximate age of the oldest known rocks, to the beginning of the Cambrian Period, roughly 540 million years ago. The Precambrian era thus represents more than 80 percent of the whole of geologic time."

15.3. 3. "It has long been known that the Cambrian marks the earliest stage in the history of the Earth when many varied forms of life evolved and were preserved extensively as fossil remains in sedimentary rocks. It is not surprising that all life-forms were long assumed to have originated in the Cambrian, and therefore all earlier rocks with no obvious fossils were grouped together into one large era, the Precambrian. However, detailed mapping and examination of Precambrian rocks on most continents have since revealed that primitive life-forms already existed more than 3.5 billion years ago. The original terminology to distinguish Precambrian from all younger rocks, nevertheless, is still used for subdividing geologic time."[2]

15.3. 4. "During the Late Proterozoic stromatolites reached their peak of development and became distributed worldwide. The first metazoa (multicelled organisms whose cells are differentiated into tissues and organs) also appeared at this time. The stromatolites diversified into complex, branching forms. From about 700 million years ago, however, they began to decline significantly in number. Possibly the newly arrived metazoa ate the stromatolitic algae, and their profuse growth destroyed the habitats of the latter."[3]

15.3. 5. Evidence mentioned above [15.3. 1–15.3. 4] shows that the presence of only primitive life forms were around 678 million years ago. As a result no life form from that time is noticeable today by the naked eye, unlike the creatures of first circle (226 million years ago). That causes no evidence about the presence of prehistoric primitive organisms in any pond.

15.3. 6. Generally humans are able to observe creatures that existed around the second SGY as fishlike creatures who lived about 452 million years ago. Any younger life forms are hardly noticeable without direct investigation. Hence, any earlier circle of SGY has no chance to bring any noticeable observation of a strange creature.

15.3. 7. However it's possible to calculate the maximum number of SGY to cover all the history of biological life on the Earth. To find that number, we need to divide the

---

[1] "Proterozoic Eon." Encyclopaedia Britannica. Encyclopaedia Britannica 2008 Deluxe Edition. Chicago: Encyclopaedia Britannica, 2008.
[2] "Precambrian time." Encyclopaedia Britannica. Encyclopaedia Britannica 2008 Deluxe Edition. Chicago: Encyclopaedia Britannica, 2008.
[3] "Precambrian time." Encyclopaedia Britannica. Encyclopaedia Britannica 2008 Deluxe Edition. Chicago: Encyclopaedia Britannica, 2008.

183

earliest estimation of biological life by the period of SGY. The calculation has the following result

$$3,500,000,000 / 226,000,000 = 3.5 \times 10^9 / 2.26 \times 10^8 = 1.55 \times 10 \approx 16 \qquad \text{(a)}$$

Hence, the Sun had about 16 full SGY in the time of an inhabited Earth, from primitive organisms to modern life forms. In the calculation we used information from [15.3. 3]: "detailed mapping and examination of Precambrian rocks on most continents have since revealed that primitive life-forms already existed more than 3.5 billion years ago".

15.3. 8. From 16 SGY, only two points from the two last SGY are observable by the Dragon Effect, as explained at [s.15.3. 6].

15.4 The Green Belt

15.4. 1. Because the Earth moves around the Sun by an elliptical trajectory, and the Sun uses the same type of orbit moving around the galaxy's center, the Sun has some deviation in distance from the galaxy center during the whole SGY. The Earth has the same deviation because its motion is bound to the Sun's location. Therefore, during the whole SGY the Earth and the Sun change a number of points with different values of galaxy SGF. In other words the S_g of the galaxy changes in some extent from a minimum to a maximum each SGY. That deviation coincides with some space in the galaxy disk (covering the whole thickness of the disk), shaped as (more or less) a thin layer surrounding the galaxy center. The figure shows that layer in a cross-section of the galaxy. Obviously, a cross-section of the layer lies inside the thicker part of the galaxy.

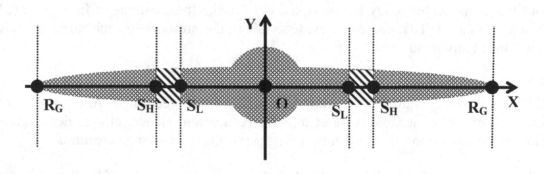

Fig. A

Figure [A] shows the cross-section of the galaxy with two axes, X and Y. Point O represents the galaxy center and point of origin for axes of reference. The figure has a symmetry relative to both axes because of the galaxy's natural shape. R_G is the most distant points from the galaxy center. Distance OR_G equals 72,000 light-years [s.15.1. 4]. Points S_L and S_H represent the highest and lowest distances between the Sun and the galaxy center during SGY. The current distance from the Sun and the galaxy center (27,000 light-years [s.15.1. 4]) is more than OS_L and less than OS_H. In other words, the location of the Sun is always between S_L and S_H.

15.4. 2. Hence, all space of the galaxy disk lies between distance OS_H and OS_L and has a value of S_g that always equals the same value of S_g at least at one location of the Sun during the full SGY. In other words for each point of the Sun's location (in SGY), a countless number of points exist with equal value of S_g located in the galaxy disc with the same value of S_g. Those points form a layer equidistant from the galaxy center. The number of such layers forms a special space with the following characteristics. Each point of that space follows equation [5.1. 10.a] and as a result gives the possibility of the existence of ZTHP between each point of that space and some locations of the Sun. Here and later I refer to such a space as the Green Belt (GB).

15.4. 3. To meet the requirements of equation [5.1. 10.a], two points of GB must have the same value of S_g. That can be reached by an exact location of the Sun and the Earth relative to the galaxy center. Each time that those locations take place, S_g has the same value and a ZTHP can possibly appear, surrounding the Sun and the Earth as its satellite.

15.4. 4. That causes the possibility for equation [5.1. 10.a] to have a solution in space and time. We can see the result of that solution just as the Dragon Effect shows to us presence of strange, prehistoric monsters. Hence, the possibility of the existence for Z-Trajectory in a large time interval (according to an IB's TKD) is caused by the Sun's motion around the galactic center.

15.4. 5. It's time to go a bit further. The solution of equation [5.1. 10.a] according to a different location of the Sun and the Earth and its consequences is explained above in details. Now there is a more complex question. Is it possible to have a different solution to that equation? Certainly it's quite possible. If anywhere in the GB there exists a star with a satellite planet, and the masses of those bodies are equal to the masses of the Sun and the Earth, the equation [5.1. 10.a] has a solution for such celestial bodies as well as for the Sun and the Earth.

15.4. 6. The only difference of that solution is this: The time interval between ZTHP calculated by an IB's TKD would be independent from the exact location of the first or second system (star-planet system) because a different system reaches points of equal S_g according to their own trajectories. As a result the time interval between two possible existences of ZTHP according to the IB's TKD has a correlation to the orbits of both systems.

15.4. 7. As a result such star-planet systems that lie on the GB can be used by Z-Process as corresponding systems for S-Gaps. Here and later I refer to such systems as Corresponding Star-Planet Systems (CSPS).

15.5 Z-Process in the Galaxy

15.5. 1. The notion of Green Belt describes the space in part of the galaxy where Z-Process is possible. As a result any number of CSPS can be used as possible location of S-Gaps. What's more, CSPS can be looked at as different locations of the same star-planet system in space and time, as described above for the Earth-Sun system, it and leads to phenomena separated by long periods of time according to the IB's TKD.

15.5. 2. In the case of the presence of CSPS at the same time according to a IB's TKD, Z-Process can use those systems as well as the same system in different points of time. Generally Z-Process itself has no difference between its presence on a single system case and different systems case.

15.5. 3. **Example**. Suppose there is a system looking like the Sun-Earth system at any point of the Green Belt. In that case there are at least two different locations reached by those systems at two different points of time. Those space-time points are possible locations for ZTHP, and Z-Process is possible between them according to equation [5.1. 10.a]. That coincides with the law of location for any other ZTHP (Z-Process is possible only between them). In that case the transposition of any moving system between two Z-Gaps would be equal to the distance between the ZTHP in space. In the area of time, transposition would be equal to the time difference between points of time when each system reaches the appropriate ZTHP according to the IB's TKD.

15.5. 4. **Example, simplified.** Suppose there is a CSPS for the Sun-Earth system. Suppose the Sun-Earth system is located at the point of a possible ZTHP presence. Suppose the corresponding ZTHP is located 10 parsecs away in a Green Belt according to the current location of the Sun-Earth system and will be reached by CSPS within 30 million years. Suppose the Z-Process begins at present time above the Earth surface (a seagull passes through the Out-Gap).

15.5. 5. In that case the seagull reaches the In-Gap of CSPS after 30 million years (according to the IB's TKD), covering a distance (from the present location of Sun-Earth system) of 10 parsecs, and finds itself above the surface of a planet from CSPS. For the seagull itself the transposition time can be calculated as ZT-Time.

15.5. 6. Hence, the SGY used previously and the transposition between time points with a distance of 226 minion years is not more than a particular case of Z-Process in the galaxy. General cases include space-time transposition between any two points belonging to Green Belt with equal value of S_g [r.5.1. 10.a].

15.5. 7. Such a possibility leads to the probable transposition of matter (rocks, solid fragments, etc.) or a BS not just between time points of a single planet and between different time points of different CSPS.

15.5. 8. Such a case coincides with the idea of panspermia. "Toward the end of the 19th century Hypothesis 3[1] gained currency, particularly with the suggestion by a Swedish chemist, S. A. Arrhenius, that life on Earth arose from panspermia, microorganisms or spores wafted through space by radiation pressure from planet to planet or solar system to solar system. Such an idea of course avoids rather than solves the problem of the origin of life. In addition, it is extremely unlikely that any microorganism could be transported by radiation pressure to the Earth over interstellar distances without being killed by the combined effects of cold, vacuum, and radiation."[2]

15.5. 9. Obviously, Mr. Arrhenius had no idea about Z-Theory. But he had the right direction of his thoughts. His theory was unacceptable because "it is extremely unlikely that any microorganism could be transported by radiation pressure to the Earth over interstellar distances without being killed by the combined effects of cold, vacuum, and radiation" [s.15.5. 8]. But Z-Theory gives the possibility of such a transfer (transposition) by means of Z-Process.

15.5. 10. In that case the presence of life at any planet, especially in the form of self-propelling organisms, leads to the possible expansion of life forms through the Green Belt. Moreover, most aspects of that process must have the result of unifying the life forms biologically, at the cellular level.

15.5. 11. Looking at the previous example, the seagull that appears on a different planet can be the first live creature on that world. Possibly it is unable to survive because of the different environment, but bacteria from its body can possibly begin a new evolution because they need different requirements from the environment to survive.

15.5. 12. Another possible way for panspermia is the transposition (by Z-Process) of any rock with bacteria imbedded on its surface. That is the realization of Mr. Arrhenius's idea about bacteria transferring from the surface of one planet to another. If that ever happened to the Earth, all biological forms of our plant are children of the Z-Process.

15.5. 13. Moreover, according to the Z-Process's affection for the Green Belt, we have the following equation about the probability of a live presence on any planet located in Green Belt.

$$P_L = P_P + \sum_{i=1}^{n} (P_i^Z) \qquad \text{(a)}$$

In the equation P_L is probability of life presence on any particular planet in Green Belt, P_P is probability of life origin at the planet, and P_i^Z is probability of transposition of any life form from planet P_i to a particular planet. The equation has the following meaning.

---

[1] The hypothesis of panspermia.
[2] "Life." Encyclopaedia Britannica. Encyclopaedia Britannica 2008 Deluxe Edition. Chicago: Encyclopaedia Britannica, 2008.

The Probability of the presence of life on any planet of Green Belt equals the sum of the originated life probability at the planet and the sum of the possible transposition of life forms by Z-Process from planet P_i to the given planet.

15.5. 14. Because Z-Process in the galaxy can exist as long as the galaxy exists itself, it's quite possible to see the result of interplanetary relocation of living creatures. Generally any evidence about the observation of a creature that cannot be identified by fossils (has no predecessor in geological time) can be supposed to be a creature that successfully passed through an interplanetary Z-Gate.

15.5. 15. The main interplanetary relationship by Z-Process in the Green Belt can be shown graphically.

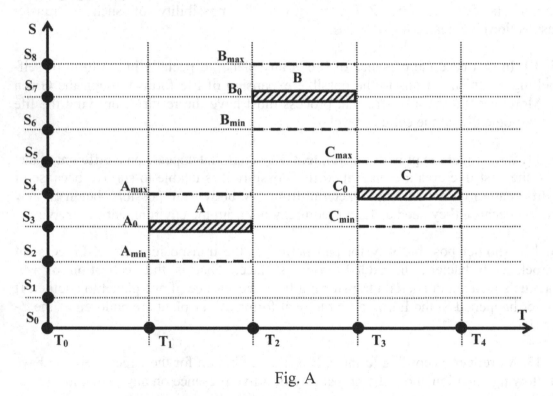

Fig. A

The figure shows three planets with different masses A, B, and C. Each planet is drawn as a rectangle with the appropriate letter above. Mass difference of the planets cause different values of S_g at the same distance from the planet center, which is mentioned at the location of the planets on the figure. Hence, planet A produces SGF with value S_3 at distance R from its center, planet B produces SGF with value S_7, and planet C produces SGF with value S_4 at the same distance from the planet center.

15.5. 16. Just like the Earth, the planets have a deviation of S_g along their surface. Maximal and minimal values of those deviations are drawn as broken lines above and below the planet S_g accordingly. Hence, the minimal possible level of S_g at the surface of planet A equals S_2 (and mentioned as A_{min}), and the maximal possible level of S_g equals S_4 (and is mentioned as A_{max}). The average value of S_g equals S_3, as already

188

mentioned, and it is marked as A_0. Other planets have the same meaning for their abbreviations.

15.5. 17. The horizontal axis is the axis of time. It shows time intervals between which the SGF of each planet has exact values. For example, the minimal value of S_g at distance R from the planet center of planet C equals S_3 between time points T_3 and T_4.

15.5. 18. All planets belong to solar systems with equal masses of the stars.

15.5. 19. In that case Z-Process is possible between any two points located at distance R from the planet center with equal value of S_g [r.5.1. 10.a]. Hence, the transposition by Z-Process is possible between planet A and C in any area where the value of S_g lies between S_3 and S_4. That happens because S_g deviations produce the same values of S_g at different areas of the planets.

15.5. 20. In the case of the presence of S_g deviation areas, the planets can have some mass deviation. For example, the more massive planet can be used as the corresponding one for Z-Process in areas where it has a value of S_g equal to some area of a different planet. That significantly increases the number of possible corresponding planets.

15.5. 21. Looking back at descriptions of Z-Sectors and the events that occur there, we can see the same meaning of figure [15.5. 15.A] and any figures mentioned relative to Z-sectors. There we have the same meaning for possible corresponding points of Z-Process as for points with exact equality of S_g caused by S_g deviations at the surface of the Earth. That is the particular case when a galaxy Z-Process involves the same planet with the same mass. In general cases the masses of corresponding planets can be more different with the greater value of gravitational anomalies that exist at the planets.

15.5. 22. Planet B [s. 15.5. 15.A] has no corresponding planets because there is no planet with the same value of S_g despite gravity anomalies of the planet leading to the S_g deviation of S_6 to S_8. Generally that planet has no chance to be involved in a Z-Process between planets A and C despite its location in the same Green Belt.

15.5. 23. Give the number of planets located in the Green Belt, their interaction by Z-Process, especially in the area of time, can be explained by the following figure.

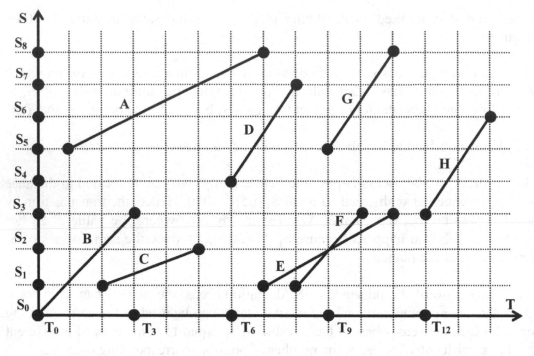

Fig. A

The figure generally has the same meaning as figure [15.5. 15.A] with two differences. It shows no S_g deviation for the planets and uses a much longer period of time in abscissa (axis 'T'). In that case the deviation of planetary masses becomes significant because of the different processes that lead to an increasing of S_g at the same distance (R) from the planet center with time. For example, young planets increase their masses (and S_g as well) because of the formation process. Elder planets can increase their masses by the natural processes of dust and meteorite absorption, especially if the planet, unlike the Erath, is located in a dusty area or in an area with a lot of asteroids.

15.5. 24. As a result any point located at the exact distance from the planet center (R) has an increasing S_g in time. That process can be shown in the Strength-Time diagram as angled lines, which represent different planets. Because the mass difference between them is quite large, no gravity anomaly can compensate that difference. So S_g at the diagram is drawn as lines related only to time and planetary masses. In other words, Z-Process is unable to exist between such planets due to the gravity anomalies (unlike figure [15.5. 15.A]).

15.5. 25. In that case each planet has its own way of planetary evolution in the Green Belt. But as soon as equation [5.1. 10.a] becomes applicable to any two or more planets (because they reach the appropriate mass and S_g), Z-Process becomes possible between those particular planets.

15.5. 26. Figure [15.5. 23] shows the graphic solution for equation [5.1. 10.a] with equal value of S_g (in same or different time). For example, if the value of S_g equals S_6, there are four planets that produce the same value of S_g at a different time. Those are planet A at time T_3, planet D at some point of time between T_7 and T_8, planet G at some point of time between T_9 and T_{10}, and planet H at time T_{14}. All points of time mentioned in the figure mean the time calculated by an IB's TKD (the bystander who is not involved in the Z-Process).

15.5. 27. As we can see there is no time point when any number of those planets produces a value of S_g equal to S_6 simultaneously. Nevertheless equation [5.1. 10.a] has a solution in the area of time. That is the primary cause for that equation to be solved mostly by time (combination of t_1 and t_2) instead of locations only (combination of x_1, y_1, z_1 and x_2, y_2, z_2). Moreover, in the Green Belt the locations related to a combination of (X,Y,Z) are negligible relative to space-time locations, each of which can have a time solution for [5.1. 10.a]. In other words the probability of a solution for [5.1. 10.a] involving time difference ($t_2 \neq t_1$) is much greater than only space solutions (in the case of $t_2 = t_1$).

15.5. 28. The only example of the [5.1. 10.a] equation for a space-only solution is the combination of planets E and F [s.15.5. 23]. Those planets produce the same value of S_g (equal to S_2) at the same distance (R) from their centers at the same time (time point T_9). In such a case the equation [5.1. 10.a] can be solved using only space solution ($t_2 = t_1$). That solution means that there are two planets in the Green Belt with equal S_g existing simultaneously. Therefore, the Z-Process between such planets is possible as a space-only transposition.

15.5. 29. Two other planets, B and C, are an additional example of corresponding planets. Z-Process is possible between them from S_1 to S_2. As we can see the time intervals for that values of S_g is different for different planets. According to planet B, that time interval has its place within unit time ($T_2 - T_1$).

15.5. 30. The time interval between S_1 and S_2 for planet C takes much more time and is equal to three time units ($T_5 - T_2$). As a result the time interval of planet B has corresponding points at three greater time intervals. What does it mean?

15.5. 31. Suppose a creature from planet C is relocated by Z-Process to planet B at the moment of time when both planets have S_g equal to S_1. That is time point T_1 for planet B and time point T_2 for planet C. The next time a Z-Process occurs at the moment of time when both planets have S_g equal to S_2, and the same type of creature relocates from planet C to planet B. That is time point T_2 for planet B and time point T_5 for planet C.

15.5. 32. In that case the IB from planet B was very surprised because of the huge biological progress of the new creature. That happens because the biological systems of planet B has had only one third the time to develop themselves, unlike the creature from corresponding planet C.

15.6 The Galaxy Belts

15.6. 1. Previous topics explained in detail some consequences that take place in the Green Belt of a galaxy. The same considerations are applicable for any galaxy and any star located in a galaxy because of the universal power of the gravity field.

15.6. 2. In the case of the Sun, we have a Green Belt, but the same considerations are applicable to any star in the galaxy and for any other solar system (combination of a star and at least one planet). Moreover, those stars can be located closer to the galaxy center or be more distant from it (though still inside the Green Belt).

15.6. 3. For each star, its own area is populated with different stars and planets which form the same structure as for the Sun. The only difference between those areas is the average distance from the galaxy center.

15.6. 4. As a result there are a lot of "belts" with any particular distance from the galaxy center. Here and later I refer to them as color belts. I refer to belts with less distance from the galaxy center than Green Belt as Red Belts (RB). For belts with an average distance from the galaxy center that exceeds the value of Green Belt, I refer to them as Violet Belts (VB). That depends on the value of S_g at different distances from the galaxy center. The closer the point is to the galaxy center, the less value of S_g it possess.

15.6. 5. Generally that happens because any point located in a galaxy disc has two opposite gravity force directions, which are produced by different parts of the galaxy. The closer the point is to a galaxy center, the less value of observable galaxy gravity attraction that exists at that point as well as S_g.[1]

15.6. 6. The figure shows the disposition of different color belts in a galaxy.

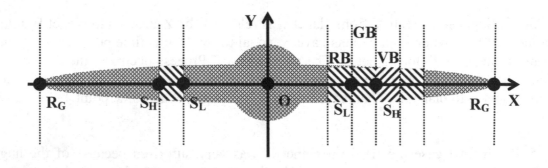

Fig. A

Abbreviations mentioned in the figure have following meaning: GB is Green Belt, RB is Red Belt, and VB is Violet Belt.

---

[1] I use the reference to the visible spectrum, where waves with different colors have different locations.

15.6. 7. Therefore, each aspect of Z-Process that can appear in a Green Belt has the possibility to appear in any different belt. Hence, if Z-Process leads to a life expansion in a Green Belt, the same process leads to the same result in any other belt. Therefore, the galaxy must be much more inhabited than we can imagine today.

15.6. 8. What's more, each belt has a number of similar stars and planets that possibly produces creatures with close characteristics (at least without much difference) because of similar masses of the corresponding stars and planets.

15.6. 9. **Example**. In the case of Earth's evolution, changing of the environment from water to dry land resulted in a range of creatures from weightless living in water to weighted living in an environment with an acceleration of g = 9,8 [r.3.1. 2]. The result of that adaptation we can see today as diversity of land animals.

15.6. 10. What happened if the Earth's mass was a few times higher or lower? Obviously, land animals and any other land life form would possess different characteristics from present-day species due to the different acceleration of free falling (because the value is proportional to planet mass).

15.6. 11. Summarizing the ideas mentioned above, we have the following consequence. Any life form that experiences an interplanetary Z-Process appears at a planet that has a different environment but the same gravity field.

15.7 Z-Process and Solar Systems

15.7. 1. It's time to discuss more the serious question about the possible maximum size of a body that can be transferred by Z-Process from one location to another. That question has a strong relation to the mass ratio between the body producing the gravity field and the object (in terminology of this work [r.3.2. 6]) using the Z-Gate.

15.7. 2. In cases of objects with negligible masses comparative to the planet mass, the planet produces the necessary gravity field by itself. If Z-Process tries to appear for more massive objects, those objects must be put in an appropriate field of a much more massive body. If that law is broken, equation [5.1. 10.a] is broken as well due to the gravity field of the object. As a result Z-Process becomes impossible.

15.7. 3. Helpfully the universe puts the planets around much more massive bodies, forming solar systems. In those systems the main body is a star with a mass incompatible with the planet masses (even put together). That star produces a strong gravity field that stabilizes the whole solar system, reaches each object, and holds it at the appropriate orbit. According to modern astronomy we have the following point of view about our own solar system.

15.7. 4. "Solar system is assemblage consisting of the Sun – an average star in the Milky Way Galaxy – and those bodies orbiting around it: 8 (formerly 9) planets with about 160 known planetary satellites (moons); countless asteroids, some with their own satellites; comets and other icy bodies; and vast reaches of highly tenuous gas and dust known as the interplanetary medium."[1]

15.7. 5. In the case of a solar system, the planet can use the star from its solar system as the Host Body [s.7.2. 5]. That leads to the possibility for the whole planet to use a Z-Gate. Moreover, the planet and any objects related to it (rocks on the planet surface, the oceans, flying birds, etc.) passes through the Z-Gate as a solid body because of the planet's gravity field.

15.7. 6. That field produces a force that refers to anything around the planet and the planet itself. To break that relationship, anything that tries to do so needs to use additional energy. That energy cannot be created from nowhere (the law of conservation restriction). As a result the Z-Process is unable (by itself) to separate any object from the transferring planet, from the tiniest atmospheric molecules to the heaviest rocks lying on the ocean bed. All those objects are bound to the planet by planetary gravitation. Even the atmosphere belongs to the planet due to the necessary mass, which helps planets hold their atmosphere. That has a strong relation to the value of escape velocity.

15.7. 7. "Escape velocity in astronomy and space exploration, the velocity that is sufficient for a body to escape from a gravitational centre of attraction without undergoing any further acceleration. Escape velocity decreases with altitude and is equal to the square root of 2 (or about 1.414) times the velocity necessary to maintain a circular orbit at the same altitude. At the surface of the Earth, if atmospheric resistance could be disregarded, escape velocity would be about 11.2 km (6.96 miles) per second. The velocity of escape from the less massive Moon is about 2.4 km per second at its surface. A planet (or satellite) cannot long retain an atmosphere if the planet's escape velocity is low enough to be near the average velocity of the gas molecules making up the atmosphere."[2]

15.7. 8. As well as gas molecules, no object has any chance to leave the planet permanently if it moves with a speed less than the escape velocity (11.3 kilometers per second for the Earth). According to Z-Process any object moving with a velocity relative to the Earth but less than escape velocity will pass through the Z-Gate with the planet.

15.7. 9. How long can the Z-Process be for whole planet? Generally ZT-Time for a planet can be calculated just as for any other object using Z-Trajectory (by equation [11.2. 21.b]). The only difference is variation of relative velocity between the planet

---

[1] "Solar system." Encyclopaedia Britannica. Encyclopaedia Britannica 2008 Deluxe Edition. Chicago: Encyclopaedia Britannica, 2008.
[2] "Escape velocity." Encyclopaedia Britannica. Encyclopaedia Britannica 2008 Deluxe Edition. Chicago: Encyclopaedia Britannica, 2008.

and the S-Gap. In the case when a S-Gap is located motionlessly relative to a star, the velocity between the planet and S-Gap can be equal to the orbital velocity of the planet. It's quite possible to calculate ZT-Time for an Earth-like planet using well-known characteristics of the planet mentioned below in the quotation.

15.7. 10. "The mean distance of Earth from the Sun is about 150 million km (93 million miles). The planet orbits the Sun in a path that is presently more nearly a circle (less eccentric) than are the orbits of all but two of the other planets, Venus and Neptune. Earth makes one revolution, or one complete orbit of the Sun, in about 365.25 days. The direction of revolution – counterclockwise as viewed down from the north – is in the same sense, or direction, as the rotation of the Sun; Earth's spin, or rotation about its axis, is also in the same sense, which is called direct or prograde. The rotation period, or length of a sidereal day (see day and sidereal time) – 23 hours, 56 minutes, and 4 seconds – is similar to that of Mars. Jupiter and most asteroids have days less than half as long, while Mercury and Venus have days more nearly comparable to their orbital periods. The 23.5° tilt, or inclination, of Earth's axis to its orbital plane, also typical, results in greater heating and more hours of daylight in one hemisphere or the other over the course of a year and so is responsible for the cyclic change of seasons.

15.7. 11. "With an equatorial radius of 6,378 km (3,963 miles), Earth is the largest of the four inner, terrestrial (rocky) planets, but it is considerably smaller than the gas giants of the outer solar system."[1]

15.7. 12. First of all we need to calculate the average orbital velocity of the Earth. That can be done using the length of the orbit (L_O) and the time of a full orbital revolution (one year; T_Y).

$$V_E = L_O / T_Y = 2 \cdot \pi \cdot r / T_Y = 2 \cdot 3.14 \cdot 150 \cdot 10^6 \text{ [km]} / 365 \cdot 24 \cdot 60 \cdot 60 \text{ [sec]} =$$
$$= 942 \cdot 10^6 \text{ [km]} / 31,536,000 \text{ [sec]} = 29.87 \approx 30 \text{ [km/s]} \quad \text{(a)}$$

Now we can calculate ZT-Time by [11.2. 21.b] using the diameter (radius multiplied by 2) of the planet as the maximum size of moving object.

$$T_Z = 2 \cdot L / V = 2 \cdot 2 \cdot 6,378 \text{ [km]} / 29.87 \text{ [km/s]} = 25,512 / 29.87 =$$
$$= 854 \text{ [sec]} = 14 \text{ min } 14 \text{ sec} \quad \text{(b)}$$

Hence, for an Earth-like planet with the same size and speed, the whole Z-Process takes place in about 15 minutes.

15.7. 13. In cases when the S-Gap positioned itself motionlessly to the frame associated with the galaxy center, the relative velocity increases because of the star's motion around the galaxy center. As one might remember, the speed of the Sun around the

[1] "Earth." Encyclopaedia Britannica. Encyclopaedia Britannica 2008 Deluxe Edition. Chicago: Encyclopaedia Britannica, 2008.

galaxy center equals 225 km/sec [r.15.1. 6]. Using the same equation for ZT-Time, we have following result.

$$T_Z = 2 \cdot L / V = 2 \cdot 2 \cdot 6,378 \text{ [km]} / (29.87 + 225) \text{ [km/s]} = 25,512 / 254.87 =$$
$$= 100.1 \text{ [sec]} = 1 \text{ min } 40 \text{ sec} \qquad \text{(a)}$$

In such a case the Z-Process for whole planet appears as a Galaxy Spin Effect (GSE) and equals (in observing characteristics) the Spin Effect for objects that are located at the Earth's surface [s.12.9. 11]. As we can see ZT-Time is very short that way.

15.7. 14. The result of such a process would appear as a matter exchange around stars (planetary exchanges) involving different stars with equal masses that are located in the Green Belt. The next topic discusses that question in detail.

15.8 Z-Process around the Sun

15.8. 1. In the case of a planetary exchange and the existence of a solar system that can be used as a corresponding one, the IB can see some unexplainable phenomena in the solar system that cannot be explained by any other theories. Because Z-Process on a solar system scale is rare, the IB can see only a few examples of that process, which took place in the past (according to the IB's TKD).

15.8. 2. The first "unexplainable" example is the Venus phenomenon. "Venus is second planet from the Sun and sixth in the solar system in size and mass. No planet approaches closer to Earth than Venus; at its nearest it is the closest large body to Earth other than the Moon. Because Venus's orbit is nearer the Sun than Earth's, the planet is always roughly in the same direction in the sky as the Sun and can be seen only in the hours near sunrise or sunset. When it is visible, it is the most brilliant planet in the sky."[1]

15.8. 3. "The rotation of Venus on its axis is unusual in both its direction and its speed. The Sun and most of the planets in the solar system rotate in a counterclockwise direction when viewed from above their north poles; this direction is called direct, or prograde. Venus, however, rotates in the opposite, or retrograde, direction. Were it not for the planet's clouds, an observer on Venus's surface would see the Sun rise in the west and set in the east. Venus spins very slowly, taking about 243 Earth days to complete one rotation with respect to the stars – the length of its sidereal day. Venus's spin and orbital periods are very nearly synchronized with Earth's orbit such that, when the two planets are at their closest, Venus presents almost the same face toward Earth. The reasons for this are complex and have to do with the gravitational interactions of Venus, Earth, and the Sun, as well as the effects of Venus's massive rotating atmosphere. Because Venus's spin axis is tilted only about 3° toward the plane of its

---

[1] "Venus." Encyclopaedia Britannica. Encyclopaedia Britannica 2008 Deluxe Edition. Chicago: Encyclopaedia Britannica, 2008.

orbit, the planet does not have appreciable seasons. Astronomers as yet have no satisfactory explanation for Venus's peculiar rotational characteristics. The idea cited most often is that, when Venus was forming from the accretion of planetary building blocks (planetesimals), one of the largest of these bodies collided with the proto-Venus in such a way as to tip it over and possibly slow its spin as well."[1]

15.8. 4. "Venus has the most massive atmosphere of the terrestrial planets, which includes Mercury, Earth, and Mars. Its gaseous envelope is composed of more than 96 percent carbon dioxide and 3.5 percent molecular nitrogen. Trace amounts of other gases are present, including carbon monoxide, sulfur dioxide, water vapour, argon, and helium. The atmospheric pressure at the planet's surface varies with surface elevation; at the elevation of the planet's mean radius it is about 95 bars, or 95 times the atmospheric pressure at Earth's surface. This is the same pressure found at a depth of about 1 km (0.6 mile) in Earth's oceans."[2]

15.8. 5. As we can see the planet Venus seems unlike any inner planet (Mercury, Earth, or Mars [s.15.8. 4]) in the areas of rotation and atmosphere. Moreover, the rotation itself shows two anomalies: angular speed and direction. That increases the number of planetary anomalies to three. That planet crushes any theory that tries to explain planetary behavior based on an idea about the natural process of formation of that planet as part of our solar system. The answer can be reached using Z-Process for the explanation.

15.8. 6. Suppose there is a planetary system in the Green Belt moving around a star with a mass equal to the mass of the Sun (CSPS). Suppose that system had the same evolution process as the planets of our system, with the only difference being the initial momentum of the whole system, which was clockwise [s.15.8. 3]. In that case the planetary transposition leads to two different possibilities explained by figure [15.8. 7.A].

15.8. 7. The following figure represents the location of two CSPS located in the Green Belt. Because of the huge radius of the galaxy, they are drawn as two systems that lie on one line (with equal value to Sg according to the galaxy)

[1] "Venus." Encyclopaedia Britannica. Encyclopaedia Britannica 2008 Deluxe Edition. Chicago: Encyclopaedia Britannica, 2008.
[2] "Venus." Encyclopaedia Britannica. Encyclopaedia Britannica 2008 Deluxe Edition. Chicago: Encyclopaedia Britannica, 2008.

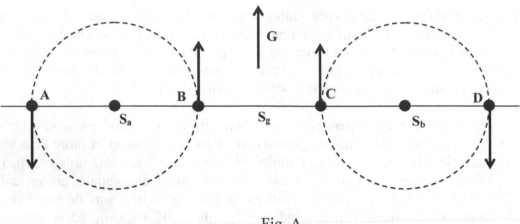

Fig. A

Generally that line can be understood as a line with equal distance from the galaxy center. Vector G shows the direction to the galaxy center.

15.8. 8. Two CSPS are represented as stars S_a (the Sun) and S_b (the star with corresponding planetary system) and two planets with orbits around the stars. The first planet belongs to the Sun's solar system and moves around the Sun in a counterclockwise direction as with the other bodies of the solar system. Direction of that motion is shown as velocity vectors from points A and B.

15.8. 9. "In astronomy, actual or apparent motion of a body in a direction opposite to that of the (direct) motions of most members of the solar system or of other astronomical systems with a preferred direction of motion. As viewed from a position in space north of the solar system (from some great distance above the Earth's North Pole), all the major planets revolve counterclockwise around the Sun, and all but Venus and Uranus rotate counterclockwise on their own axes; these two, therefore, have retrograde motion. Of the known satellites of the planets, a minority display retrograde revolution. These include the four outermost moons of Jupiter; Phoebe, the outermost moon of Saturn; and Triton, the largest of Neptune's moons. The orbital planes of the satellites of Uranus are tilted so greatly that the description of these bodies' motion as either retrograde or direct has little meaning. The revolutions around the Sun of all known asteroids are direct; of the known periodic comets, only Halley's Comet moves in a retrograde orbit."[1]

15.8. 10. Unlike a solar system forming around the Sun, in the corresponding system the natural motion is clockwise (because of its initial momentum difference). As a result most members of the corresponding system rotate clockwise around their star.

15.8. 11. According to the relative location of both systems to the galaxy, there are four points lying on orbits of corresponding planets with equal values of S_g marked as A,

---

[1] "Retrograde motion." Encyclopaedia Britannica. Encyclopaedia Britannica 2008 Deluxe Edition. Chicago: Encyclopaedia Britannica, 2008.

B,C, and D. Those points can be used as ZTHP [r.7.1. 27]. Z-Process itself can appear as a transposition between points A-D and B-C [b.7.1].

15.8. 12. As we can see, the velocity vectors of both systems coincide at those points. Therefore, when a planet is relocated from point C (for example) to point B, the relocation keeps its energy unchanged both ways via linear motion and rotation (a law of conservation restriction). As a result, after Z-Gate the planet moving with its usual velocity keeps the same orbit around a corresponding star because the distance between the planet and corresponding star is the same as before the Z-Process. Everything is best that way, but according to the law of conservation the planetary rotation keeps unchanged as well as the speed of the planet before and after Z-Gate. Hence, the IB watching such a planet is able to see only an "unnatural" revolution of the planet around its own axis.

15.8. 13. It is only possible way for Z-Process because the transposition between points A-C and B-D, where velocity vectors of the planets are opposite to each other, is restricted by the law of conservation [b.7.1]. It's related to star-planet locations, not to the directions of planet orbital velocity.

15.8. 14. As well as the planet with the unusual revolution around the axis, its atmosphere relocates with the planet [b.15.7. 7]. That is the answer to the question about the density and unusual atmospheric pressure of Venus [s.15.8. 4].

15.8. 15. Any theory that tries to explain the retrograde motion of Venus is refuted by the law of conservation. That happens because the revolving planet can be imagined as an enormous gyroscope with a great amount of energy related to the revolution around the axis. As a result anybody that tries to change the direction of the planet axis and its revolution must have an appropriate mass and velocity to hit the planet and withstand the revolving energy of the whole planet. Such an impact would destroy the planet (and the colliding body) completely instead of changing the direction of the revolution or the axis direction.

15.8. 16. The Venus example appears in cases of the absence of a planet in the orbit of the corresponding star. What happens if the relocating planet meets a planet in the corresponding system with an orbit similar in characteristics with its own one? That leads to the presence of two planets in two close orbits. As a result those planets have a strong gravity attraction the closer to each other they are during their orbital motion. Because of the close distance between their orbits, the two planets sooner or later crash into each other.

15.8. 17. The result of such a catastrophe would be a lot of fragments from both planets keeping orbit with characteristics not far from the orbits of the planets before the collision. What's more the mass of the whole number of fragments and the direction of their motion around the star would be the same as the motion of the planets (a law of conservation restriction). Sooner or later the fragments would spread around the star, filling the whole orbit because of the gravity disturbance from other planets and the

199

slightly different velocity (comparative to each other) of the fragments from the initial impact.

15.8. 18. In our solar system, such an area filled by a lot of asteroids exists between the orbits of Mars and Jupiter. Usually they call that area the asteroid belt. "Asteroid also called minor planet or planetoid any of a host of rocky small bodies, about 1,000 km (600 miles) or less in diameter, that orbit the Sun primarily between the orbits of Mars and Jupiter in a nearly flat ring called the asteroid belt. It is because of their small size and large numbers relative to the major planets that asteroids are also called minor planets."[1]

15.8. 19. Suppose the masses of the planets did not have a significant difference. In that case if the planets had opposite orbital directions before collision, most parts of the kinetic energy must be compensated at the time of collision (the law of conservation). As a result most parts of fragments after the collision did not have enough velocity to keep a planetlike orbit. The Sun, accelerating such fragments, changed their trajectories to either an extended elliptical orbit or a trajectory with a final point on the Sun's surface. That does not coincide with present-day observation of the asteroid belt.

15.8. 20. As we can see, Z-Theory gives an alternative explanation of the asteroid belt origin as well as for Venus's retrograde motion. Modern astronomy has the following point of view about the asteroid belt.

15.8. 21. "Available evidence indicates that the asteroids are the remnants of a 'stillborn' planet. It is thought that at the time the planets were forming from the low-velocity collisions among asteroid-size planetesimals, one of them grew at a high rate and to a size larger than the others. In the final stages of its formation this planet, Jupiter, gravitationally scattered large planetesimals, some of which may have been as massive as Earth is today. These planetesimals were eventually either captured by Jupiter or another of the giant planets or ejected from the solar system. While they were passing through the inner solar system, however, such large planetesimals strongly perturbed the orbits of the planetesimals in the region of the asteroid belt, raising their mutual velocities to the average 5 km per second they exhibit today. The increased velocities ended the accretionary collisions in this region by transforming them into catastrophic disruptions. Only objects larger than about 500 km in diameter could have survived collisions with objects of comparable size at collision velocities of 5 km per second. Since that time, the asteroids have been collisionally evolving so that, with the exception of the very largest, most present-day asteroids are either remnants or fragments of past collisions."[2]

---

[1] "Asteroid." Encyclopaedia Britannica. Encyclopaedia Britannica 2008 Deluxe Edition. Chicago: Encyclopaedia Britannica, 2008.
[2] "Asteroid." Encyclopaedia Britannica. Encyclopaedia Britannica 2008 Deluxe Edition. Chicago: Encyclopaedia Britannica, 2008.

15.9 Z-Process and SGY Trace

15.9. 1. Previous topics discussed cases of Z-Process for whole planets. It's time to think about cases when Z-Process involves only planet fragments instead of a whole planet. In other words this topic discusses traces of events that are possible as a result of Z-Process between CSPS on the scale of planet fragments (e.g., rocks of various size and other pieces of matter transferring from the planet to a corresponding one, and vice versa).

15.9. 2. First there must be information about some type of rock that does not belong to the Earth and is recognized by a number of differences from usual rock. The more differences it has, the more probable its origin from a different planet. The most likely candidate for such an object is the object with abbreviation ALH84001.

15.9. 3. "ALH84001 meteorite determined to have come from Mars and the subject of a contentious scientific claim that it contains the remains of ancient life indigenous to the planet. Recovered from the Allan Hills ice field of Antarctica in 1984, the 1.9-kg (4.2-pound) igneous rock is thought to have crystallized from magma on Mars 4.5 billion years ago and later to have been shocked and altered, perhaps by one or more nearby meteoroid or asteroid impacts. Still later, carbonate mineral grains were introduced into shock-induced fractures in the rock. Another large impact blasted the rock off the Martian surface and into solar orbit with a velocity greater than the planet's escape velocity of 5 km per second (11,000 miles per hour). Over time the rock's orbit was altered such that it approached Earth, eventually falling in Antarctica about 13,000 years ago.

15.9. 4. "In 1996 NASA scientists who carried out microscopic and chemical analyses on ALH84001 touched off a controversy by suggesting that the carbonates in the meteorite had been produced by Martian microorganisms. The carbonate grains are associated with organic compounds, contain minute crystals of iron minerals similar in size and shape to those produced by bacteria, and exhibit elongated objects resembling microscopic fossils. In subsequent investigations, other scientists contested the interpretation of these lines of evidence, demonstrating that each could be adequately explained by nonbiological processes or was not entirely consistent with what is known about microfossils and living microorganisms on Earth. The hypothesis that ALH84001 contains evidence for extraterrestrial life has not found wide acceptance, although there are strong indications that Mars may once have been hospitable to life."[1]

15.9. 5. Additionally, there is some other evidence about meteorites from the Mars. "By 2004 scientists had identified about 30 meteorites that have come from Mars. Suspicions about their origin were first raised when meteorites that appeared to be volcanic rocks were found to have ages of about 1.3 billion years instead of the 4.5 billion years of all other meteorites. These rocks had to have come from a body that

---

[1] "ALH84001." Encyclopaedia Britannica. <u>Encyclopaedia Britannica 2008 Deluxe Edition</u>. Chicago: Encyclopaedia Britannica, 2008.

was geologically active in the comparatively recent past, and Mars was the most likely candidate. The rocks also have similar ratios of oxygen isotopes, which are distinctively different from those of Earth rocks, lunar rocks, and other meteorites. A Martian origin was finally proved when it was found that several of them contained trapped gases having a composition identical to that of the Martian atmosphere as measured by the Viking landers. The rocks are thought to have been ejected from the Martian surface by large impacts. They then went into solar orbit for several million years before falling on Earth. Claims in the mid-1990s of finding evidence for past microscopic life in one of the meteorites, called ALH84001, have been viewed skeptically by the general science community."[1]

15.9. 6. Summarizing the information mentioned above, we have the following evidence.

   a. There are some rocks
   b. Those rocks were recognized as objects with extraterrestrial origin
   c. One rock contains something that was recognized as "evidence for past microscopic life" [s.15.9. 5]
   d. Several rocks "contained trapped gases" [s.15.9. 5]

15.9. 7. Obviously, the general science community gave an explanation of the rocks' origin according to the nearest planet from which the meteorites had came from. However most of them disagree with the idea of marks of life found on one rock. That happened because the force of impact on Mars' surface by an asteroid that could accelerate the rocks to escape velocity would erase any evidence of life. Such an amount of energy always leads to a melting of rocks. Hence, they transform to new objects in which only chemical characteristics coincide with the original planet surface. The following quotation shows that process in detail.

15.9. 8. "When an asteroidal or cometary object strikes a planetary surface, it is traveling typically at several tens of kilometres per second – many times the speed of sound. A collision at such extreme speeds is called a hypervelocity impact. Although the resulting depression may bear some resemblance to the hole that results from throwing a pebble into a sandbox, the physical process that occurs is actually much closer to that of an atomic bomb explosion. A large meteorite impact releases an enormous amount of kinetic energy in a small area over a short time. Planetary scientists' knowledge of the crater-formation process is derived from field studies of nuclear and chemical explosions and of rocket missile impacts, from laboratory simulations of impacts using gun-impelled high-velocity projectiles, from computer models of the sequence of crater formation, and from observations of meteorite craters themselves."[2]

---

[1] "Mars." Encyclopaedia Britannica. Encyclopaedia Britannica 2008 Deluxe Edition. Chicago: Encyclopaedia Britannica, 2008.
[2] "Meteorite crater." Encyclopaedia Britannica. Encyclopaedia Britannica 2008 Deluxe Edition. Chicago: Encyclopaedia Britannica, 2008.

15.9. 9. Meteorite theory used for the explanation of those rocks' presence is only acceptable for traditional science. But it's quite possible to have an alternative explanation using Z-Process.

15.9. 10. First of all I'd like to turn our attention to the map [appendix C] that was used for explanations related to Z-Sectors. The area mentioned as a native one for the meteorites is the "Allan Hills ice field of Antarctica" [s.15.9. 3]. According to the map, that area coincides with Z-Sector (F), "the large area located south-east from Australia (the continental part of Antarctica) and off the coastal area" [r.14.3. 2.f]. The area named "Victoria Land" is located there.

15.9. 11. "Victoria Land is physical region in eastern Antarctica, bounded by the Ross Sea (east) and Wilkes Land (west) and lying north of the Ross Ice Shelf. It was discovered in 1841 by a British expedition led by Sir James Clark Ross, and it was named for Queen Victoria. It consists largely of snow-covered mountains, with heights up to 13,668 feet (4,166 metres) and a network of outlet glaciers draining the adjacent East Antarctic ice sheet. Major ecological and paleoclimate studies have been made in the dry valleys of Victoria Land east of McMurdo Sound.

15.9. 12. "The United States and New Zealand operate research stations there. More than 300 meteorites preserved in Antarctic ice were located in 1979 in the Allan Hills and Darwin Glacier areas of Victoria Land."[1]

15.9. 13. The next quotation gives us a detailed description of Antarctic meteorites and their origin area. "Antarctic meteorite is any of a large group of meteorites that have been collected in Antarctica, first by Japanese expeditions and subsequently by U.S. and European teams since the discovery of meteorite concentrations there in 1969. Although meteorites fall more or less uniformly over Earth's surface, many that fall in Antarctica are frozen into its ice sheets, which slowly flow from the centre of the continent toward its edges. In some places, patches of ice become stranded behind mountain peaks and are forced to flow upward. These stagnant patches are eroded by strong winds, thereby exposing and concentrating meteorites on the ice surface. Such areas, called blue ice for their colour, have over just a few decades provided more than 15,000 individual meteorites ranging in size from thumbnail to basketball. Although many meteorites are paired (parts of the same original fall), the Antarctic collection still represents several thousand new samples, which is comparable to the total number of catalogued meteorites that were collected elsewhere over the past several centuries.

15.9. 14. "Because large concentrations of Antarctic meteorites occur within small areas, the traditional geographic naming system used for meteorites is not applicable. Rather, they are identified by an abbreviated name of some local landmark plus a number that identifies the year of recovery and the specific sample. For example, the meteorite ALHA81005 was found in the Allan Hills region in 1981 and is the fifth sample recovered.

---

[1] "Victoria Land." Encyclopaedia Britannica. Encyclopaedia Britannica 2008 Deluxe Edition. Chicago: Encyclopaedia Britannica, 2008.

15.9. 15. "Antarctic meteorites have provided additional specimens of poorly represented meteorite types and of a few types that were previously unknown. Meteorites from the Moon were first recognized in Antarctica, and most lunar and many Martian meteorites have been collected there. Antarctic meteorites have spent times on Earth that range from a few thousand to about a million years. They thus provide insight into the kinds and abundances of meteorites that fell to Earth before recorded history."[1]

15.9. 16. Hence, the rocks are moved by ice shields from the center of Antarctic toward its edges. The map at appendix C shows that place as the continental NDA with a deviation level equal to -60 mGal. That is the original area of all collected meteorites.

15.9. 17. Because "meteorites fall more or less uniformly over Earth's surface" [s.15.9. 13], they can be found in Antarctica as well as in any different area. But in the case of Z-Sector, any rock found at that particular area must be analyzed to ascertain the possibility of Z-Process. If any characteristics for such a meteorite has a significant difference from the usual type of objects, we have evidence for Z-Process successfully happening in the past.

15.9. 18. Moreover, only Z-Process is able to bring to another planet surface any life forms or evidence of its presence from the origin planet. Suppose scientists who believe in the presence of such evidence are right, and the rock ALH84001 really contains traces of biological life. What does it mean?

15.9. 19. First of all we need to remember the interaction between tiny forms of biological life (bacteria) and animals with their usual size. Sometimes that interaction can be positive, when both forms take advantage from the interaction and establish a biological collaboration. Usually such an interaction is called mutualism.

15.9. 20. "Mutualism is association between organisms of two different species in which each is benefited. Mutualistic arrangements are most likely to develop between organisms with widely differing living requirements. The partnership between nitrogen-fixing bacteria and leguminous plants is an example, as is the association between cows and rumen bacteria (the bacteria live in the digestive tract and help digest the plants eaten by the cow). The associations between tree roots and certain fungi are often mutualistic"[2]

15.9. 21. Other ways of biological interaction between species are parasitism and parasitoidism. "Parasitism is relationship between two species of plants or animals in which one benefits at the expense of the other, sometimes without killing it. Parasitism

---

[1] "Antarctic meteorite." Encyclopaedia Britannica. Encyclopaedia Britannica 2008 Deluxe Edition. Chicago: Encyclopaedia Britannica, 2008.
[2] "Mutualism." Encyclopaedia Britannica. Encyclopaedia Britannica 2008 Deluxe Edition. Chicago: Encyclopaedia Britannica, 2008.

is differentiated from parasitoidism, a relationship in which the host is always killed by the parasite."[1]

15.9. 22. Sometimes an unwilling intrusion of parasites and batteries causes strong diseases that can be dangerous for the host. "Disease most commonly is caused by the invasion of an organism by one or more outside agents. Typically the infectious organisms are microorganisms (e.g., bacteria, viruses, and fungi), but they also can include larger organisms such as parasitic worms or nonliving but harmful substances such as toxins or ionizing radiation."[2]

15.9. 23. Now suppose there is a planet with biological life. That life develops itself according to common evolutionary principles. That state exists as long as no changes happen.

15.9. 24. Suppose at some particular time a Z-Gate brings to the planet a rock with microorganisms from another planet. Those organisms appear in a different and unusual environment for them. Interaction between those organisms and original species of the planet can cause two different results. The species can interact positively and negatively. The first way leads to mutualism [s.15.9. 20], or sometimes to the absence of noticeable interaction between species.

15.9. 25. The second way leads to a harmful result for the original species. In that case the intruding species causes widely spread disease, killing most of the original species (on a planet scale). As a result a lot of fossils from that particular time would appear for investigators at a later geological epoch.

15.9. 26. In other words that way appears as a biological process usual to any time when any species meets a changed environment or a different species in same environment. Only species that possessed strong immunity for the new disease survived and kept evolving further. Such results show no information in fossils about causes of mass extinctions. That happens because fossils contain no information about infections or parasites on (in) the bodies of their victims. As a result modern paleontology has no strong answer to the question about causes of mass extinctions.

15.9. 27. "Best known among mass extinctions is the one that occurred at the end of the Cretaceous Period, when the dinosaurs and many other marine and land animals disappeared. Most scientists believe that the Cretaceous mass extinction was provoked by the impact of an asteroid or comet on the tip of the Yucatan Peninsula in southeastern Mexico 65 million years ago. The object's impact caused an enormous dust cloud, which greatly reduced the Sun's radiation reaching Earth, with a consequent drastic drop in temperature and other adverse conditions. Among animals, about 76 percent of species, 47 percent of genera, and 16 percent of families became extinct.

---

[1] "Parasitism." Encyclopaedia Britannica. Encyclopaedia Britannica 2008 Deluxe Edition. Chicago: Encyclopaedia Britannica, 2008.
[2] "Disease." Encyclopaedia Britannica. Encyclopaedia Britannica 2008 Deluxe Edition. Chicago: Encyclopaedia Britannica, 2008.

Although the dinosaurs vanished, turtles, snakes, lizards, crocodiles, and other reptiles, as well as some mammals and birds, survived. Mammals that lived prior to the event were small and mostly nocturnal, but during the ensuing Tertiary Period they experienced an explosive diversification in size and morphology, occupying ecological niches vacated by the dinosaurs. Most of the orders and families of mammals now in existence originated in the first 10 million – 20 million years after the dinosaurs' extinction. Birds also greatly diversified at that time."[1]

15.9. 28. "Several other mass extinctions have occurred since the Cambrian. The most catastrophic happened at the end of the Permian Period, about 248 million years ago, when 95 percent of species, 82 percent of genera, and 51 percent of families of animals became extinct. (See also Triassic Period: Permian-Triassic extinctions.) Other large mass extinctions occurred at or near the end of the Ordovician (about 440 million years ago, 85 percent of species extinct), Devonian (about 360 million years ago, 83 percent of species extinct), and Triassic (about 210 million years ago, 80 percent of species extinct). Changes of climate and chemical composition of the atmosphere appear to have caused these mass extinctions; there is no convincing evidence that they resulted from cosmic impacts. Like other mass extinctions, they were followed by the origin or rapid diversification of various kinds of organisms. The first mammals and dinosaurs appeared after the late Permian extinction, and the first vascular plants after the Late Ordovician extinction."[2]

15.9. 29. Information about mass extinctions can be represented in the form of a table with extinction name and time. The table shows that aggregated information sorted by time.

Table A

Mass extinction Name	Time of Mass Extinction
Cretaceous	65 million years ago
Triassic	210 million years ago
Permian	248 million years ago
Devonian	360 million years ago
Ordovician	440 million years ago

First glance brings no useful information from that table. But in conjunction with the notion of SGY, we have a very interesting figure that is presented by appendix D with the name "Sun Galaxy Year and Mass Extinctions".

15.9. 30. That figure shows SGY on two axes, X and Y. Suppose the observer looks at the galaxy from the point when the Sun (as well as the planets around it)moves around the galaxy center in a counterclockwise direction.

---

[1] "Evolution." Encyclopaedia Britannica. Encyclopaedia Britannica 2008 Deluxe Edition. Chicago: Encyclopaedia Britannica, 2008.
[2] "Evolution." Encyclopaedia Britannica. Encyclopaedia Britannica 2008 Deluxe Edition. Chicago: Encyclopaedia Britannica, 2008.

15.9. 31. In that case negative time points associated with the past would appear at a clockwise direction around the point of origin. The present point of time is shown as the point that belongs to the positive direction of abscissa (point P).

15.9. 32. Hence, the whole Sun Galaxy Year can be represented as a circle around the point of origin. As mentioned at [15.1. 12], the full period of a Sun Galaxy Year equals 226 million years. As a result that point coincides with the present time in the figure.

15.9. 33. The inner circle represents points of time in the Sun Galaxy Year and the location of the Sun (and planetary system) around the galaxy center in the past. For example, at half the SGY period of time before the present time, the Sun was located at the opposite point relative to the galaxy center and was positioned at the point with mark 113 (half time of SGY). That means 169.5 million years before present time, the Sun took a position that coincides with the positive direction of the Y axis. That is three-quarters of the full SGY.

15.9. 34. Going further in the past, the Sun took a different location according to the time difference between the present point of time and a past one. For example, 448 million years before present time, the Sun was located at the same point as today relative to the galaxy center, because that time coincides with two full SGY (point P).

15.9. 35. The outer circle (appendix D) shows specific points of time according to their location on a SGY. Those points coincide with the time of well-known mass extinctions. As a result we have the following points. Point A is the Cretaceous mass extinction (65 million years ago), point B is the Triassic mass extinction (210 million years ago), point C is the Permian mass extinction (248 million years ago), point D is the Devonian mass extinction (360 million years ago), and point E is the Ordovician mass extinction (440 million years ago).

15.9. 36. According to the SGY, period we have the following distance between points of present time and points of mass extinctions.

Tab. A

Mass Extinction Name	Time of Mass Extinction (in millions of years)	Relation to SGY	Point in Figure (Appendix D)
Cretaceous	65	113 - 48	A
Triassic	210	226 - 16	B
Permian	248	226 + 22	C
Devonian	360	226 + 134 = 226 + 113 + 21	D
Ordovician	440	226·2 - 12	E

The table shows how far the sun was from its present location when the mass extinctions happened. For example, the Triassic mass extinction has a result of 226 -

16. That means the Sun needs 16 million additional years to reach the same point (relative to the galaxy center) as it had at the time of the Triassic mass extinction. The next example is the Permian mass extinction, whose calculation has a result of 226 + 22. That mean this 22 million years ago the Sun had passed (relative to the galaxy center) the point that it had at the time of the Permian mass extinction (point C).

15.9. 37. One might have already noticed a strange closeness of most points of mass extinctions to the present location of the Sun (points B, C, and E). The distance between them and the present location of the Sun is much less than the distance from points A and D. Generally points B, C, and E occupy part of the Sun's trajectory that is covered by the star within 38 million years. That is calculable from a maximum distance (in time) between the most distanced points B (-16 million years) and C (+ 22 million years). That is 16.8 per cent of the SGY whole period.

15.9. 38. The location of these mass extinction points can be easily explained by means of the Sun's location according to the its trajectory around the galaxy center. In the case of an elliptical trajectory, the distance between the galaxy center and the Sun has little difference from point to point. According to figure [10.2. 1.A], for each point of elliptical trajectory (except aphelion and perihelion), there are two points with equal value of S_g. Those points can be used as points of an S-Outline (ZTHP) [s.10.2. 8].

15.9. 39. Suppose points B and C are different points of the same S-Outline. In that case the Sun's orbit around the galaxy center must reach aphelion or perihelion between those points in a location equidistant from both points [r.10.2. 1.A]. The exact value can be reached by dividing the time between those points (38 million years [s.15.9. 37]) by two (period of time to reaching aphelion or perihelion).

$$38,000,000 \text{ [years]} / 2 = 19,000,000 \text{ [years]} \tag{a}$$

The calculation gives a result of 19 million years. That is the time distance between each of those points and the aphelion or perihelion. In that case the present location of the Sun relative to aphelion or perihelion can be calculated as the time difference between location B (or C) and the present location. Using the information of point B, we have the following result.

$$(P-AP_1) = (B-AP_1) - (B-P) = 19 - 16 = 3 \text{ [Million years]} \tag{b}$$

In the figure of appendix D, AP_1 is point of aphelion or perihelion, $P-AP_1$ is distance between present location of the Sun and its aphelion or perihelion, $B-AP_1$ is distance between point of Triassic mass extinction and Sun aphelion or perihelion, and B-P is distance between present location of the Sun and point of Triassic mass extinction. That result has following meaning.

15.9. 40. The Sun's present location is distanced from the point of aphelion or perihelion by 3 million years. In other words the Sun passed the point of aphelion or perihelion 3 million years ago.

15.9. 41. Hence, the solar orbit according to the galaxy appears with the following parameters. The sun moves around the galaxy center by an elliptical trajectory. Each full revolution around the galaxy center (SGY) takes 226 million years. The present location of the Sun is distanced from the nearest point of the line of apsides [r.10.2. 3][1] for 3 million years. The Sun has already passed that point (3 million years ago).

15.9. 42. To simplify the following explanations, I refer point to AP_1 as the point of aphelion (Sun is at the farthest retreat from the galaxy center). That is possible because of the elliptical trajectory's symmetry. If further investigations give the opposite result, the point of AP_1 can be used as the point of perihelion because according to the purpose of this work, the exact distance between each location of the Sun and the Galaxy center is not important. The only important thing is the location of points with equal value of S_g. As a result any consideration of using two points with equal S_g located closer to perihelion is applicable for two different points located closer to aphelion. The only difference would be in the radius of the circle with an equal value of S_g [r. 10.2. 1.A].

15.9. 43. Using that consideration, we have the following consequences: the line of apsides is located between points AP_1 and AP_2. Moreover, point AP_1 turns to aphelion, and the opposite point (AP_2) turns to perihelion.

15.9. 44. It's quite easy to recalculate the mass extinction point's location using the distance between the current location of the Sun and the aphelion (3 million years ago) relatively to the Sun's galaxy orbit. The table shows the result of those calculations.

Table A

Mass Extinction Name	Time of mass extinction (in millions of years)	Relation to SGY	Point in Figure (Appendix D)	Distance from Aphelion (in millions of years)
Cretaceous	65	113 - 48	A	113 - 51
Triassic	210	226 - 16	B	-19
Permian	248	226 + 22	C	+19
Devonian	360	226 + 113 + 21	D	113 + 19
Ordovician	440	226·2 - 12	E	-9

15.9. 45. The following figure shows the information from table [15.9. 44.A] graphically.

---

[1] With regard to the Sun's motion around the galaxy center.

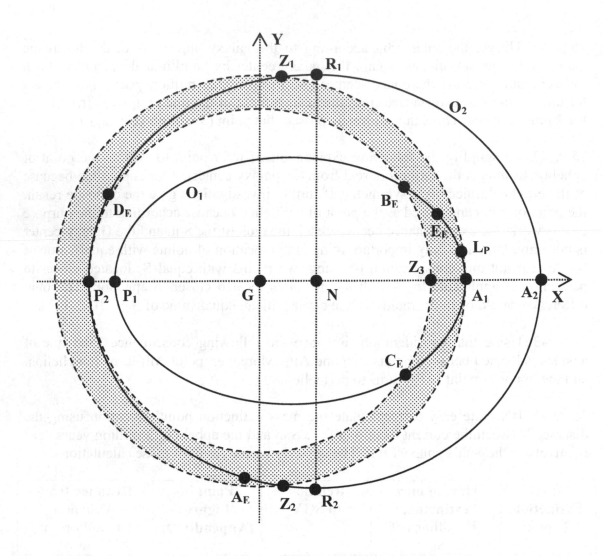

Fig. A

In figure [A] there are two drawn orbits of CSPS O_1 and O_2 and their relation by Z-Process. Both orbits have the same focus, G, that represents the galaxy center. The orbits have an axis of upside that coincides with the X axis of the figure. P_1 and P_2 are points of perihelion of those orbits just as A_1 and A_2 are points of aphelion. R_1-R_2 is the little axis of the O_2 orbit (ellipse) that crosses the X axis at point N. Generally the figure represents two CSPS using two very close orbits in a galaxy.

15.9. 46. Suppose the Sun uses O_1 orbit and CSPS uses O_2 orbit. Because of the elliptical orbits, the perihelion of each orbit has less distance from the galaxy center than the aphelion.[1] Hence, the perihelion distances are GP_1 and GP_2, and the aphelion distances are GA_1 and GA_2. In case the aphelion of orbit O_1 has a greater distance from the galaxy center than the perihelion of orbit O_2, a little layer appears between them as a number of points equidistant from the galaxy center. Parts of both CSPS orbits crossing that layer exist as a number of ZTHP. Z-Process is possible between any pair

---

[1] This is the law of any elliptical trajectory.

of those points equidistant from the galaxy center [r.5.1. 10.a]. Here and later I refer to that area as the Z-Ring. The Z-Ring has the following notion.

15.9. 47. Z-Ring is between two close stars' elliptical orbits with a different semimajor axis from the galaxy center, and it appears in the galaxy as a layer limited by the distance of the perihelion of the orbit with a greater semimajor axis and the aphelion of the orbit with a lesser semimajor axis.

15.9. 48. As we can see in figure [15.9. 45.A], a Z-Ring is only possible if the perihelion of the orbit with the greater semimajor axis lies closer to the galaxy center than the aphelion of the orbit with the lesser semimajor axis. Z-Ring is shown in figure [15.9. 45.A] as a dotted ring between two broken lines representing a number of points with equal values of S_g. For example, S_g has the same value for the following pairs of points: P_2-Z_3, Z_2-A_1, C_E-B_E, et cetera.

15.9. 49. In that case any point of orbit O_2 at location Z_1-P_2-Z_2 has at least one equivalent point (ZTHP) at orbit O_1 at location C_E-A_1-B_E. In other words Z-Process is possible between orbits parts Z_1-P_2-Z_2 and C_E-A_1-B_E [r.5.1. 10.a]. Generally the figure [15.9. 45.A] shows a graphical solution for that equation on a galaxy scale.

15.9. 50. Now suppose the existence of two CSPS using orbits O_1 and O_2 [s.15.9. 45.A]. In that case he Z-Process described at [15.9. 23–15.9. 28] is possible between them. As a result, from time to time an interaction between species from different planets causes a mass extinction at one planet. That process is caused mostly by bacterial and viral diseases.

15.9. 51. To spread such disease, the virus or bacteria must survive in the environment of the new planet and infect the original planetary species. Those combinations cause very unusual epidemics because the immune system of original species has no power against an infection that it has never meet before.

15.9. 52. "From the lowliest protozoans to the higher marine tunicates, invertebrates have means of distinguishing self components from nonself components. Sponges from one colony will reject tissue grafts from a different colony but will accept grafts from their own. When tissue grafts are made in animals higher up the evolutionary tree – between individual annelid worms or starfish, for example – the foreign tissue is commonly invaded by phagocytic cells (cells that engulf and destroy foreign material) and cells resembling lymphocytes (white blood cells of the immune system), and it is destroyed. Yet tissues grafted from one part of the body to another on the same individual adhere and heal readily and remain healthy. So it seems that something akin to cellular immunity is present at this level of evolution."[1]

15.9. 53. "Insects engulf and eliminate foreign invaders through the process of phagocytosis ('cellular eating'). They have factors present in their circulatory fluids

[1] "Immune system." Encyclopaedia Britannica. Encyclopaedia Britannica 2008 Deluxe Edition. Chicago: Encyclopaedia Britannica, 2008.

that can bind to foreign cells and cause clumping, or agglutination, of a number of these cells, an event that facilitates phagocytosis. Insects also seem to acquire immunity to infectious agents."[1]

15.9. 54. "The most sophisticated immune systems are those of the vertebrates. Recognizable lymphocytes and immunoglobulins (Ig; also called antibodies) appear only in these organisms. The most primitive living vertebrates – the jawless fishes (hagfish and lampreys) – do not have lymphoid tissues corresponding to a spleen or a thymus, and their immune responses, although demonstrable, are very weak and sluggish. Farther up the evolutionary tree, at the level of the cartilaginous fishes (sharks and rays) and the bony fishes, a thymus and a spleen are present, as are immunoglobulins, although only those immunoglobulins of the IgM class are detectable. Fish lack specialized lymph nodes, but they do have clusters of lymphocytes in the gut that may serve an analogous purpose."[2]

15.9. 55. In the case of the Earth's concentration of mass extinctions points relative to the Sun orbit, they form an area that includes the Triassic mass extinction, the Permian mass extinction, and the Ordovician mass extinction (points B_E, C_E, and E_E in figure [15.9. 45.A]). Moreover, points B_E and C_E were used to determinate the most distant points of the Z-Ring from the aphelion. For the two last SGY there were no mass extinctions with a greater distance (in time) between the Sun's location at aphelion and the point of mass extinctions.

15.9. 56. As we can seen in figure [15.9. 45.A], points P_2, C_E, and B_E belong to the same galaxy S-Outline. That leads to the following conclusion. Those mass extinctions can be caused by consequences of the same Z-Process happening between those points. In other words the Z-Process between points P_2-B_E causes the same result (mass extinction) as for P_2-C_E. Points B_E, C_E, and E_E coincide with points (B, C, and E) in the Sun's galaxy year, drawn in appendix D.

15.9. 57. Point E_E (he Ordovician mass extinction) is located in the same Z-Ring as the Triassic and Permian mass extinctions. What's more, the current position of the Sun is located in same Z-Ring. As a result the same process is possible now. Is there any evidence of that process? There is the best kind of evidence of that process: the object ALH84001. That is an example of something that happened many times in past. But the relocation to Antarctica causes with high probability a freezing of any living bacteria and virus. Hence, that is the example and answer to the question as to why mass extinctions are not a permanent process [r.15.9. 51].

15.9. 58. According to Z-Process, the probability of such epidemics exceeds zero only in part of the whole trajectory that belongs to a Z-Ring.

---

[1] "Immune system." Encyclopaedia Britannica. Encyclopaedia Britannica 2008 Deluxe Edition. Chicago: Encyclopaedia Britannica, 2008.
[2] "Immune system." Encyclopaedia Britannica. Encyclopaedia Britannica 2008 Deluxe Edition. Chicago: Encyclopaedia Britannica, 2008.

15.9. 59. In other parts of trajectory, the probability of such epidemics can be calculated by following equation.

$$P_E = P_G \cdot P_P \cdot P_S \cdot P_C \qquad \text{(a)}$$

In the equation P_E is probability of mass extinction, P_G is probability of Z-Gate appearance, P_P is probability of successful passing of the Z-Gate by bacteria or viruses, P_S is probability of intrusion of microorganisms surviving in a new environment, and P_C is probability of contagion (spreading disease among original planetary species).

15.9. 60. In cases where star location is out of Z-Ring, P_G and P_E become zero. In the case of the Antarctic, P_S becomes almost zero, and P_E takes the same value.

15.9. 61. Looking back to [15.9. 4], we can see an excellent example from the ALH84001 description. "In subsequent investigations, other scientists contested the interpretation of these lines of evidence, demonstrating that each could be adequately explained by nonbiological processes or was not entirely consistent with *what is known about microfossils and living microorganisms on Earth.*"

15.9. 62. That is true because there are no living microorganisms on Earth that are able to produce Earth's biology, unlike the evidence of their existence. In other words, only biology unlike Earth's microorganisms (CSPS, extraterrestrial) can produce evidence of their presence. The process of their relocation (Z-Process) happens permanently as soon as the Sun takes position inside the Z-Ring.

15.9. 63. Just as the Sun can have CSPS in a galaxy orbit with a greater semimajor axis, it can have CSPS with a lesser semimajor axis (that is located closer to the galaxy center). In that case figure [s.15.9. 45.A] is still useful with only one difference: the Sun uses orbit O_2. As a result CSPS associates with orbit O_1.

15.9. 64. Such a relation between orbits of two CSPS leads to the existence of Z-Ring between them. According to the Sun's orbit (O_2), the Z-Ring begins at point Z_1, goes through point P_2, and ends at point Z_2.

15.9. 65. As we can see from figure [15.9. 45.A], points Z_1 and Z_2 are closer to the perihelion (P_2) than to the aphelion (A_2). That happens because of the elliptical nature of the orbits (O_1, O_2), their coinciding focuses at the galaxy center (G), and different semimajor axes.

15.9. 66. Part of trajectory Z_1-P_2-Z_2 coincides with the second Z-Ring (the lesser one). To separate both Z-Rings, here and later I refer the Z-Ring with a higher average radius as Violet-Z-Ring (V-ZR), and the Z-Ring with a lower average radius as Red-Z-Ring (R-ZR).

15.9. 67. Anything mentioned above according to V-ZR is fully applicable to R-ZR. As a result that area of the Sun's orbit can have points of mass extinctions too. There are two examples of that process that have happened in the past. Those are the points of the Cretaceous mass extinction (point A_E) and the Devonian mass extinction (point D_E). Unlike the previously discussed mass extinctions, those extinctions were caused by infection from the CSPS of R-ZR. That infection was different from one of V-ZR infections.

15.9. 68. Summarizing the considerations mentioned above, it's quite possible to draw a relationship between SGY and both Z-Rings. The following figure shows that graphically.

In the figure (see below), SGY is drawn as an ellipse with the galaxy center at the points G (one of the ellipse focuses). Part of the Sun's trajectory belongs to the Red-Z-Ring (R-ZR) and the Violet-Z-Ring (V-ZR), marked by dotted spaces. There are four specific points in the Sun's galaxy trajectory where the trajectory meets and leaves both Z-Rings. Those are points V_1, V_2, R_1, and R_2. Points (V_1, V_2) are equidistant from the aphelion (point A) of the Sun's galaxy orbit just as points (R_1, R2) are equidistant from the perihelion (point P) of the orbit.

According to fossil information and the consideration mentioned above [s.15.9. 39], V_1 and V_2 have a time distance from the aphelion equal to 19 million years. Points R_1 and R_2 have a time distance from the perihelion equal to 51 million years (using the same type of calculation).

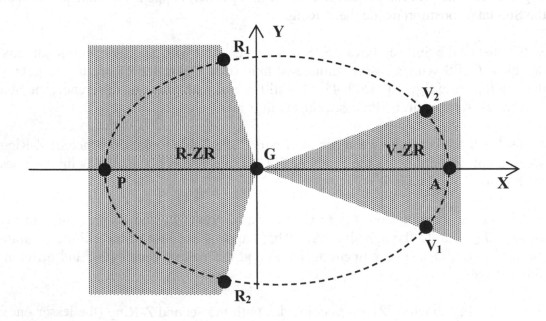

Fig. A

15.10 Z-Bridge and Z-Range

15.10. 1. Z-Rings and the Green Belt have a strong relationship due to the galaxy orbits of CSPS. The figure shows that relationship graphically.

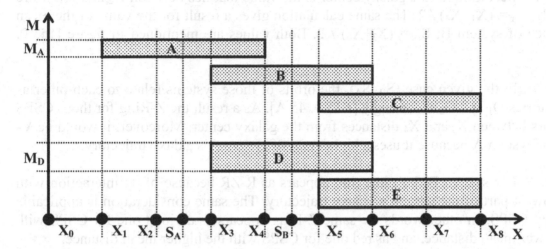

Fig. A

The figure shows two axes, X and M. Axis X shows the distance from the galaxy center, and axis M shows the relative masses of CSPS planets. Orbital parameters of the stars of planet systems are shown as a length of rectangles. For example, the perihelion distance of star A (at minimum distance between the galaxy center and the star) is X_1. The aphelion distance of the same star according to the galaxy center is X_4. Hence, star A is always located between distance X_1 and X_4 relative to the galaxy center.

The planet in the planetary system of star A has mass M_A (shown at the axis of relative masses). That mass is proportional to the height of rectangle A. The height of other rectangles has the same meaning.

15.10. 2. As we can see in figure [15.10. 1], star B from the next planetary system has distances X_3 and X_6 as its perihelion and aphelion (aspects of star trajectory are relative to the galaxy center). Masses of the planets A and B are equal to each other (shown as equal heights of rectangles A and B). As a result those star-planet systems can be used as a corresponding one.

15.10. 3. For each star, the distance between perihelion and aphelion of its galaxy trajectory is the Green Belt [s.15.4. 2] (minimum and maximum distance from the galaxy center). Hence, the Green Belt for star B lies between X_3 and X_6, for star C it lies between X_5 and X_8, et cetera.

15.10. 4. For A-B CSPS there is a distance between X_3 and X_4 where the value of S_g is equal in both systems. As a result Z-Process is possible between those CSPS at any location between X_3 and X_4. As we can see system A has a lesser semimajor axis than system B ($X_4 < X_6$). According to figure [15.10. 1.A] there is a mean distance that can be found between values of semimajor axis and minor semiaxis according to the given orbit (parameters of the appropriate rectangle). For example, the mean distance of system A according to the galaxy center is the value that lies between X_2 and X_3. More exactly, $S_a = (X_1 + X_4) / 2$. The same calculation gives a result for the value of the mean distance of system B: $S_B = (X_3 + X_6) / 2$. Both values are mentioned in figure [15.10. 1.A].

15.10. 5. In the given case ($S_B > S_A$), the orbits of those systems relate to each other as trajectories O_1 and O_2 accordingly [s.15.9. 45.A]. As a result the Z-Ring for those CSPS appears between X_3 and X_4 distances from the galaxy center. Moreover, it would be V-ZR for system A because it uses a higher part of system A's galaxy trajectory.

15.10. 6. For system B the same ring appears as R-ZR because of its interaction with the lowest part of the system's galaxy trajectory. The same consideration is applicable for any CSPS. In other words the same Z-Ring appears as a violet one for CSPS with the lower mean distance, and as red one for CSPS with the higher mean distance.

15.10. 7. For example, system B has two Z-Rings. The red Z-Ring occupies the distance between X_3 and X_4. That Z-Ring uses system A as a corresponding point and produces its influence by Z-Process between systems A and B.

The violet Z-Ring occupies the distance between X_5 and X_6. That Z-Ring uses system C as a corresponding point and produces its influence by Z-Process between systems B and C.

15.10. 8. System D cannot be used as a corresponding point for system A despite the same orbital parameters as system B (same values of perihelion and aphelion). That happens because of the different masses of the planets from systems A and D (different heights of rectangles A and D). As a result Z-Process is impossible between those star-planet systems despite the orbital possibility for a Z-Process presence (between X_3 and X_4).

15.10. 9. Unlike systems A and B, systems D and E can be used as corresponding points because of equal rings belonging to both systems located between X_5 and X_6 and equal masses of the planets. As a result the galaxy area between X_5 and X_6 becomes a Z-Ring (a violet Z-Ring for system D and a red one for system E).

15.10. 10. Suppose anything carrying any type of living species passed through a Z-Gate in the Z-Ring between CSPS A and B. Suppose the natural environment of the corresponding planet of system B is acceptable for the adaptation of a species from the planet of system A. As a result the species will spread across the galaxy from the X_1-X_4 Green Belt to the X_3-X_6 Green Belt. Therefore, that species spreads farther from the galaxy center for a X_4-X_6 distance.

15.10. 11. The next step is using a Violet Z-Ring of system B that leads a species to system C and spreading it up to X_8 distance from the galaxy center (using the Green Belt of system C). As a result the species that originated at system A is able to spread far away from its original location (the Green Belt of its star, X_1-X_4).

15.10. 12. As we can see system B helps to spread life from system A and C, which cannot be used as a CSPS directly to each other because of the absence of a Z-Ring between them. In other words, system B can be used as a bridge connecting different CSPS by its red and violet Z-Rings.

15.10. 13. Here and later I refer to the combination of a star-planet system and its Violet Z-Ring and Red Z-Ring that can be used by Z-Gate leading for two different star-planet systems as a Z-Bridge.

15.10. 14. For the combination of CSPS inside the galaxy that relates to each other by a number of Z-Bridges, I refer to them here and later as the Z-Range.

15.10. 15. In other words the Z-Range appears as the maximum possible distance between two different points relative to the galaxy center that can be covered by matter, living or inanimate, using Z-Gates; a rock can be transferred between CSPS just as well as a living creature.

15.10. 16. From a human's point of view, life is more significant than lifeless matter, but Z-Process uses both objects with or without life signs through the same physical law. Hence, life spreading through a Z-Range is only a consequence of the whole Z-Process and its aspects.

15.10. 17. For example, figure [15.10. 1.A] shows two different Z-Ranges. The first one appears as distance X_1-X_8 covered by the Green Belts of systems A, B, and C. The second one appears as distance X_3-X_8 covered by the Green Belts of the last two systems (D, E).

15.11 Life and the Galaxy

15.11. 1. Previous topics [15.9–15.10] show examples of Z-Process in Z-Rings relative to SGY. That period (described at [15.9] as a possible period of time for easily observable creature relocation) covers the last two Sun Galaxy Years (approximately 446 million years). But the presence of Z-Rings is possible just as the Sun's galaxy trajectory doesn't change. Hence, Z-Rings and their phenomena must have existed for a time that greatly exceeded a single SGY.

15.11. 2. In that case the processes of interaction between CSPS at Z-Rings happened again and again as soon as the Sun reached any of the Z-Rings. Some of those Z-Gates that can appear at each Z-Ring relocate more or less matter between planets (rocks, etc.) than others, and some relocated species.

15.11. 3. If the presence of those species has a harmful impact for the original planetary system, the new species spread diseases among them as mentioned before at [s.15.9]. That case is possible only if the species survive on the corresponding planet after the transposition (using Z-Gate). In other cases the following question arises. What happened if no original species existed on the corresponding planet at the time of species transposition (according to the IB's TKD as related to the planet)?

15.11. 4. That species becomes used to the corresponding planet as its new home, following evolution in the usual way, adapting to the environment in order to survive and reproduce. Here arises question about species. Which type of species has the maximum adaptive power to meet a changed environment and has the best chance to survive? The answer is this.

15.11. 5. "Bacteria singular bacterium any of a group of microscopic organisms that are prokaryotic, i.e., that lack a membrane-bound nucleus and organelles. They are unicellular (one-celled) and may have spherical (coccus), rodlike (bacillus), or curved (vibrio, spirillum, or spirochete) bodies."[1]

"Bacteria can be found in all natural environments, often in extremely large numbers. As a group, they display exceedingly diverse metabolic capabilities and use almost any organic compound, and even some inorganic salts, as a food source. Some bacteria cause disease in humans, animals, or plants, but most are harmless or beneficial ecological agents whose metabolic activities sustain higher life-forms. Without bacteria, soil would not be fertile, and dead organic material would decay much more slowly."[2]

"In a sense, bacteria are the dominant living creatures on Earth, having been present for perhaps three-quarters of Earth history and having adapted to almost all available ecological habitats."[3]

15.11. 6. Hence, the only type of species has "exceedingly diverse metabolic capabilities". That type is bacteria. In case of transposition of some number of species only bacteria has maximal chance to colonized corresponding planet and begin new evolution in environment of new planet.

15.11. 7. That is the indirect answer on the question of life's origin. Any evidence about life's origin on the Earth has as a start point the evidence of bacteria (not their predecessors). "Among the oldest known fossils are those found in the Fig Tree chert from the Transvaal, dated at 3,100,000,000 years old. These organisms have been

---

[1] "Bacteria." Encyclopaedia Britannica. Encyclopaedia Britannica 2008 Deluxe Edition. Chicago: Encyclopaedia Britannica, 2008.
[2] "Bacteria." Encyclopaedia Britannica. Encyclopaedia Britannica 2008 Deluxe Edition. Chicago: Encyclopaedia Britannica, 2008.
[3] "Bacteria." Encyclopaedia Britannica. Encyclopaedia Britannica 2008 Deluxe Edition. Chicago: Encyclopaedia Britannica, 2008.

identified as bacteria and blue-green algae. It is very reasonable that the oldest fossils should be procaryotes rather than eucaryotes. Even procaryotes, however, are exceedingly complicated organisms and very highly evolved. Since the Earth is about 4,500,000,000 years old, this suggests that the origin of life must have occurred within a few hundred million years of that time."[1]

15.11. 8. That is a straight description for the first step of life's evolution on Earth. That step was done by "exceedingly complicated organisms" that were "very highly evolved" [s.15.11. 7]. That happened because the same organisms were relocated by one of the Z-Gates and began Earth's colonization as "exceedingly complicated organisms". The following calculations give a more exact sense about the location of the Sun in its galaxy orbit at that time.

15.11. 9. As mentioned at [15.11. 7], the oldest known fossils are dated at 3,100,000,000 years old. That period of time takes 13.72 SGY (3,100,000,000 / 226,000,000 = 13.72). Where was the Sun on the galaxy orbit that time? The nearest full SGY is 14 SGY. Therefore, we have the time distance between the present position and the Sun's location at that time equal to following.

$$14 - 13.72 = 0.28 \qquad (a)$$

That is the difference between the present and past location of the Sun according to SGY. It's quite possible to recalculate the distance of SGY (0.28 of full SGY) to millions of years from the present location.

$$0.28 \cdot 226,000,000 \text{ [years]} = 63,280,000 \text{ [years]} \qquad (b)$$

According to the location of the Sun relative to aphelion (3 million years ago), that point took place 60.28 million years ago. That is an area of Red Z-Ring [s.15.9. 68.A] as well as the closest location of a mass extinction point (Cretaceous mass extinction). The time distance between those points relative to SGY equals 1.7 million years (65 - 63.3 = 1.7).

15.11. 10. Suppose that the mass extinction was caused by the same Z-Process as bacteria's presence 3,100,000,000 years ago. In that case the first bacteria was relocated to the Earth by Z-Process 3,100,000,000 + 1,700,000 = 3,101,700,000 years ago. They used 1.7 million years or the colonization of the Earth and were possible to detect in the fossil records of that period.

15.11. 11. That leads to the following conclusion. The predecessor species for life on Earth was relocated from Red Z-Ring 3,101,700,000 years ago. In other words, the Red Z-Ring is the "Mother Ring" for Earth life.

---

[1] "Life." Encyclopaedia Britannica. Encyclopaedia Britannica 2008 Deluxe Edition. Chicago: Encyclopaedia Britannica, 2008.

15.11. 12. According to fossil information, there is no evidence of an older presence of life (bacteria and blue-green algae) [r.15.11. 7]. That is evidence itself about the failed attempts of Earth colonization by life from any other Z-Ring before that time.

15.11. 13. After that point of time, the Z-Rings again and again try to spread life to Earth. Sometimes those attempts were dangerous, as described in detail at [15.9]. Hence, the Z-Process (in any Z-Ring) leads not only to life spreading between CSPS, but the same process leads to life destruction if new intruders (species) produce a harmful impact to species already present. Therefore, life appearance and mass extinctions are different results of the same process, which takes place in the galaxy as long as the galaxy exists (and has the same value of S_g at the same distance from the galaxy center).

15.11. 14. As a result there was not a point of time when life began. It was a permanent process that happened again and again at least a few times at each SGY while the Earth was not ready to care for life. As soon as the Earth environment became acceptable for life from CSPS, Earth's evolution began.

15.11. 15. The Earth, as well as the CSPS from both Z-Rings, is able to produce the same type of influence on life forms to corresponding planets. In that case the Z-Bridge including the Earth-Sun system was involved in spreading life from the Red Z-Ring to the Violet Z-Ring. In other words life is spreading in Earth's Z-Range from the galaxy center to the far parts of the galaxy (in the Red-Violet direction).

15.12 Buckshot Effect

15.12. 1. As mentioned in [10.2. 16], Z-Process is possible between mirror points. That is an aspect of elliptical trajectory relative to Z-Process. Generally the difference in time between those points can be calculated by [10.2. 14.a–10.2. 14.b]. The relation to that equation was used previously to determine possible mirror points in the Sun' galaxy trajectory using information of mass extinctions. As we can see, figures [10.2. 1.A] and [15.9. 45.A] have points with equal meaning. Those mirror points are B and D ([10.2. 1.A]) and B_E and C_E ([15.9. 45.A]).

15.12. 2. For such points, the equation [10.2. 14.a] is applicable (with the [10.2. 14.b] restriction, of course). As we can see from [10.2. 14], the value of n can be any integer number (without any additional restriction). That leads to the following conclusion.

15.12. 3. Any Z-Process appearing at any ZTHP can lead transferring systems, objects, et cetera to any number of points separated by time relative to the parameters of the body's elliptical trajectory. The figure shows that graphically.

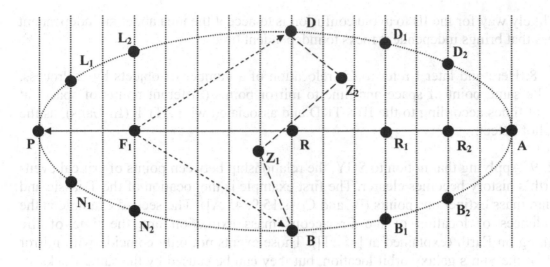

Fig. A

In the figure, the main ellipse represents the elliptical trajectory of the body moving around point F_1 as the trajectory's focus contains the central body. Points A and P are aphelion and perihelion accordingly. Other points form pairs with equal value of S_g (as equidistance from the central body) and with the same subscripts. For example, points D and B have the same value of S_g as well as the pairs D_1-B_1 and D_2-B_2. Those points are connected by vertical straight lines with a meaning of corresponding points (ZTHP) and mirror points as well. Generally the figure looks equal to [10.2. 1.A].

15.12. 4. Suppose there is a Z-Process that brings something to point B_1. The Z-Process is able to bring that same thing to a different point, D_1, because of equal value of S_g at both points. Hence, as soon as the body takes its location at point B_1 or D_1, the same result of same Z-Process is able to appear regardless of the number of orbital periods that separate those locations in time [r.10.2. 14.a]. As a result a Z-Process with one Out-Gap is able to relocate something from one point to mirror points of a different body's trajectory any number of times.

15.12. 5. Suppose the Out-Gap goes to the planet surface (by Spin Effect) covered by rocks. That Out-Gap reaches each rock at a different point of time because of the distance between them. It can be a very small time difference.

15.12. 6. In that case each rock can be relocated to the same point of space (B_1, for example) and time (t_1, for example). As a result the rocks appear at point B_1 at time t_1. But equation [10.2. 14.a] permits the rocks to use a mirror point (point D_1) as ZTHP and In-Gap location. Moreover, both points can be used with the time shift equal to a full revolution of the body around the central body.

15.12. 7. Therefore, different rocks from the same location can be relocated by Z-Gate to different points and different times. Such relocation is viewed from the IB associated to the body (a planet) as a number of independent rocks appearing from nowhere. The

221

most likely way for the IB to avoid confusion is to accept the idea about an independent process that brings independent rocks to independent times.

15.12. 8. Here and later I refer to the relocation of a number of objects by Z-Process, from the same point of space and time to mirror points (different points of space) at different times according to the IB's TKD and associated with ZTHP (In-Gaps), as the Buckshot Effect.

15.12. 9. Applying that notion to SGY, the relationship between points of critical evens of Earth's history becomes clearer. The first example is the location of the Triassic and Permian mass extinctions points (B_E and C_E [s.15.9. 45.A]). The second example is the coincidences of location of the Cretaceous mass extinction and the time of life beginning on Earth (explained at [15.11]). Those events not only coincide with mirror points of the Sun's galaxy orbit location, but they can be caused by the same Buckshot Effect. In that case the Cretaceous flora and fauna met biological intruders exactly equal to the ones that began Earth's colonization 3,101,700,000 - 65,000,000 = 3,036,700,000 years before the Cretaceous mass extinction.

15.12. 10. The Buckshot Effect is very useful to determine missing points of events. If the location of any event is known relative to the location of the Sun's galaxy orbit, it's quite possible to calculate the mirror point where the missing event is (was, will be) possible to occur. According to elliptical trajectory, a mirror point must have a location equidistant from the closest aphelion or perihelion [s.15.12. 3.A].

15.12. 11. Using that concept, we can find some missing points. The nearest point lies 9 million years before aphelion (12 million years before present time). That point is the mirror for the Ordovician mass extinction. I refer to that point as the Ordovician missing point. The Triassic and Permian mass extinctions are mirror points to each other. The Devonian mass extinction has a missing point (Devonian missing point) with location 113 - 19 + 3 = 97 million years ago ([s.15.9. 44.A; appendix D]). And the most interesting point is the missing point of the beginning of life. That point coincides with the Cretaceous missing point but is distanced from it by much more time. The Cretaceous missing point has a location 113 + 51 = 164 million years before present time. And the missing point for life beginning is 3,101,700,000 + 51,000,000 + 51,000,000 = 3,203,700,000 years ago (the location of the first point of life transposition, plus the time for perihelion, plus an equal time to the missing point (see appendix D)). That is the next most probable point of life beginning on Earth. That coincides too with [15.11. 7]: "this suggests that the origin of life must have occurred within a few hundred million years of that time".

15.12. 12. As already mentioned at [r.15.9. 59.a], the mass extinction at any point has a given probability. That is the answer to the question about the presence of missing extinction points. If circumstances are not enough for the intruding organisms to survive, mass extinction never happens. The most important evidence of that is the information of ALH84001.

15.13 "Strange" Rocks

15.13. 1. The object ALH84001 was put by Z-Process in an extremely cold environment of the planet [s.15.9. 3]. That environment was unchanged for a very long period of time. Being frozen in the Antarctic ice, the object ALH84001 had no chance to spread any life forms (even microscopic ones) in Earth's natural environment.

15.13. 2. The same result happened in the past, leading to missing points of mass extinctions. Moreover, in cases of an unacceptable environment, the life forms are unable to colonize the new planet.

15.13. 3. The most significant physical characteristic of strange rocks is their age, equal to 1.3 billion years [s.15.9. 5]. Usually such a characteristic is explained as "These rocks had to have come from a body that was geologically active in the comparatively recent past" [s.15.9. 5]. But Z-Process lets such rocks appear with different ages relative to the IB's TKD [b.12.8. 19.a].

15.13. 4. Because of the small size of the rocks, their ZT-Time appears as a very low value [r.14.1. 10.a]. As a result in the case of strange rocks, ZT-Time becomes negligible relative to altered time and transposition time. In that case the altered time and transposition time can be used as same value.

15.13. 5. The figure graphically shows the relation between altered time, transposition time, and the age of the rocks in cases when the rocks were relocated from CSPS to the planet.

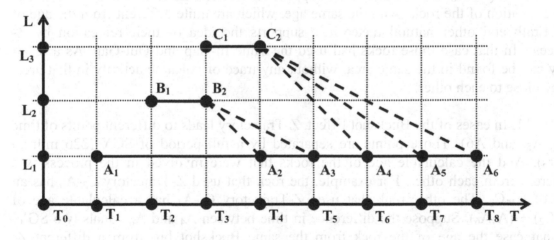

Fig. A

The figure shows the relation between locations of the "strange" rocks and the time points calculated according to the IB's TKD. There are three different locations, mentioned as L_1, L_2, and L_3. Location L_1 is associated with the Earth. Two other locations (L_2 and L_3) are associated with two different planets of CSPS (from the red and violet rings).

223

15.13. 6. The timeline associated with the IB's TKD (an IB located on the Earth, location L_1) is shown as an abscissa with time points from T_0 to T_8. A_1-A_6 is the Earth's age (4.5 billion years), A_1 is the time point of Earth's origin, and A_6 is the point of present time.

15.13. 7. B_1 and C_1 are points of time when two different planets of two different CSPS were created. According to the IB's TKD, those events happened at T_2 and T_3 points of time, respectively.

15.13. 8. Under usual circumstances the oldest rocks of the Earth have an age equal to A_1-A_6 period of time. But in the case of Z-Process, the probability of revealing rocks with differences in age raises significantly. Moreover, the presence of such rocks has no relation to volcanic activity.

15.13. 9. At T_4 moment of time, the rocks are relocated from location L_3 to location L_1 (Z-Trajectory C_2-A_6). In that case the age of the rocks after relocation can be found as equal to time interval C_1-C_2 because that is the only time when the rocks were in CE-Space (and interact with the universe in the usual way).

15.13. 10. According to information about the "meteorites from the Mars", their age is equal to 1.3 billion years [s.15.9. 5]. But any rock from the corresponding planet of CSPS is able to have the same age. In cases of Z-Process that value becomes equal to the time of the rocks' presence on their original planet. Figure [15.13. 5.A] shows that period of time as C_1-C_2. The same figure clearly shows the difference between the age of the Earth (A_1-A_6) and the age of the supposed meteorites from the Mars (C_1-C_2). The close location of the rocks with the same age, which are quite different from the age of the Erath and other natural meteorites, supports the idea of their relocation by Z-Process. In that case those rocks had used the same In-Gap and Out-Gap. As a result they can be found in the same area, without any trace of volcanic activity in that area, lying close to each other.

15.13. 11. In cases of the Buckshot Effect, Z-Trajectory leads to different points of time (A_4, A_5, and A6). Those points are separated by a full period of SGY (226 million years). And the calculable age of the rocks that were involved in the process was different from each other. For example, the rock that used Z-Trajectory C_2-A_6 has an age of C_1-C_2. The other rock that used Z-Trajectory C_2-A_4 has a calculable age of $(C_1C_2) + (A_4A_6)$. Suppose the difference in time between A_4 and A_6 equals two SGY. In that case the age of the rock from the same Buckshot but from a different Z-Trajectory (C_2A_4) possesses the following value: $1.3 + 0.226 + 0.226 = 1.752$ billion years.

15.13. 12. The same process that happened in past leads to the presence of rocks with different ages. For example, the rock that was relocated from location L_2 (red Z-Ring) by Z-Trajectory B_2A_3 with an initial age of 1.3 billion years (B_1B_2 time) has an age of $B_1B_2+A_3A_6$. If A_3A_6 period of time equals 2 billion years (for example). the full age of such a relocated rock becomes 1.3 + 2 = 3.3 billion years. As a result any relocated rock has a very high probability of having different age from nearby, usual ones, because of the difference between ZT-Time, transposition time, and altered time [r.12.8. 19.a].

In the case of the rocks, ZT-Time becomes tiny [r.14.1. 10], and altered time becomes almost equal to transposition time.

15.13. 13. If the rocks come from the same celestial body, they need to use some particular points of the body's trajectory [r.15.1]. In cases of the same point of SGY, any relocation leads to a time difference in the rocks' age equal to the number of full SGY (226 million years) [s.10.2. 12.b]. Therefore, the age difference between usual rocks and the relocated ones by Z-Process from the Earth to the Earth has the following values for five consequent circles.

Table A

SGY Number	Age Difference (in billions of years)
1	0.226
2	0.452
3	0.678
4	0.904
5	1.130

One might recognize the first and second values of the table. They were already used in relation to biological objects [r.15.2]. That happens because Z-Process uses biological and nonbiological objects in the same way as any other physical process.

15.13. 14. Table [15.13. 13.A] containing the calculation results of equation [10.2. 12.b] helps us to separate rocks by their origin. If the rock has the usual age, it must be a rock that was present on the Earth's surface as any other object. If it has an age difference equal to one of the numbers mentioned at [15.13. 13.A] or calculated by [10.2. 12.b] (for SGY), that rock must be one that was relocated by Z-Process.

15.13. 15. **Example**. Suppose someone had found a rock with an age of 3.47 billion years. The age difference calculation gives us 4.6 - 3.47 = 1.13 billion years (4.6 billion years is the age of the Earth [r.14.2. 47]). Dividing that number for SGY, we have the following result 1.13 / 0.226 = 5 (integer). Hence, that rock is the rock from the fifth circle of the Sun around the galaxy center (SGY). Such a calculation is especially true in cases of witnesses who saw "falling rocks from the sky" (not meteorites; sometimes they have the name "aerolites" in written sources). That supports evidence about the nonvolcanic origin of such rocks (see appendix B).

15.13. 16. In the case of "meteorites from the Mars" with an age of 1.3 billion years [r.15.9. 5], the age difference between them and usual rocks equals 4.6 - 1.3 = 3.3 billion years. According to SGY it contains 3.3 / 0.226 = 14.60 circles. That is not an integer value. As a result such rocks cannot be rocks from any previous location of the Sun-Earth system (the same point regarding SGY).

15.13. 17. Those rocks can be only from a CSPS. Other evidence about their appearance can be directly observed only when the Sun takes a location in the Violet-Z-Ring or Red-Z-Ring [r.15.9. 68.A]. The current location of the Sun in the galaxy orbit gives the possibility for such a process because the Sun is located in the Violet-Z-Ring (point L_P in the figure [15.9. 45.A]).

Chapter 16. Human-related Appearances

16.1 Missing Airplane Event

16.1. 1. This topic is dedicated to the explanation of some mysterious events that took place at different times across the world. The general purpose of this topic is to explain the types of events instead of each single event because Z-Theory helps us to understand those events as groups of the same events. Each type of event described below is a reconstruction of a single possible event that involved humans with the following explanation.

16.1. 2. **Reconstruction of the possible event (RPE).** Event type is a missing airplane.

16.1. 3. George Taylor and William Donovan were the first and second pilots of a flight crew. They spent a lot of hours together guiding large airplanes above the North Atlantic. Both of them were experienced pilots with an excellent reputation in company.

16.1. 4. That day was a usual flight to the United States. Heavy vehicles using the full power of jets had already crossed the ocean and were prepared to landing in the airport. Everything onboard was well. Both pilots stared through the cockpit windows, guiding the aircraft to the correct direction and watching the angle, speed, and a dozen other indications from the electronic equipment.

16.1. 5. Landing clearance was already received from the control tower, and the pilots were ready to see the runway from far away. Low-level clouds blocked a clear vision of some directions, but that didn't disturb the pilots because the clouds were no matter for their powerful aircraft.

16.1. 6. Suddenly one of the clouds appeared directly before the airplane [s.16.1. 10.a] during its reduction of altitude. Both pilots did nothing to avoid the cloud. They were sure the cloud would be penetrated by the aircraft as well as a lot the other clouds before. A moment later the aircraft was fully swallowed by the cloud [s.16.1. 10.b].

16.1. 7. A white shroud covered the cockpit windows for a couple of seconds as the aircraft flew inside the cloud [s.16.1. 10.a]. George glanced at the indications of the control panel. All indications showed a straight path to the airport [s.16.1. 10.c].

16.1. 8. A second later the aircraft had left the cloud, and the pilots restored their clear vision of the surrounding space [s.16.1. 10.d]. Immediately they heard a worried dispatch asking about their condition [s.16.1. 10.e]. William answered the request and asked about the situation. The dispatcher report confused both pilots. The person said they had been absent for about a quarter of an hour [s.16.1. 10.f].

16.1. 9. Both pilots checked all indications of the control panel but saw nothing unusual [s.16.1. 10.g]. They reported that to the dispatcher and confirmed their intention to land [s.16.1. 10.h].

16.1. 10. Explanation of the [16.1. 2] event type.

a. The Out (In)-Gap was mistaken as a cloud. For more details, see [12.1] (cloud mistake).
b. To pass a Z-Gate, the object must go through the Out-Gap completely. For more details, see [11.2].
c. Because of the high velocity of the aircraft, most indicators were unable to change indications noticeably for a short period of time (less than one second or so). For more details, see [12.8] (ZT-Time calculation for an aircraft, especially [12.8. 13]).
d. Because of the high velocity of the aircraft, ZT-Time was very short. For more details, see [12.8. 13].
e. Radio communication was established again as soon as the aircraft left ARI (Gap-Hood). The pilots had no chance to notice the break in radio communication because of the very small value of ZT-Time (just as they were unable to notice a difference in indications of the control panel at that time).
f. The dispatcher acted as an IB with an independent TKD. As a result the difference between ZT-Time and altered time appears obvious as soon as the aircraft came back to CE-Space. For more details, see [12.8].
g. After Z-Process any physical process behaved in its usual way. As a result all indications showed nothing unusual.
h. Human-related decision: if there is nothing unusual onboard, it's quite possible to do everything the usual way.

16.2 Missing Crew Event

16.2. 1. The most doomed event associated with ships was the event of a missing crew. Sometimes a ship finds another ship without any sign of its crew. They produced a lot of theories that attempted to explain such mysterious disappearances. Z-Theory gives an answer to that process as well as a description of more interesting events happening across the world. Hence, the missing crew is the only side of Z-Process that involves humans. That is its most important aspect.

16.2. 2. **Reconstruction of possible event (RPE)**. Event type is a missing crew.

16.2. 3. Robert Brown was a real sea dog. He spent many years on the open sea and became a fearless and resolute captain who took command of a large cargo ship a few years ago. Since that time he had been the only captain of the ship. The crew respected him for his honor and professional skill.

16.2. 4. One day the captain was woken up by his first mate, Joseph Jefferson, before the usual time. The worried first mate told him something that was hard to believe. Within a couple of minutes both men were at the pilot house. All crew members were already on the deck. The captain glanced around and felt frozen blood in his veins.

16.2. 5. The entire sky from horizon to horizon had a terrible green coloration [b.16.2. 17.a]. No onboard man ever seen a stranger natural event. Green color dominated everywhere, from the sky to the water. The hole body of the ship looked deep green instead of white, its usual color.

16.2. 6. Suddenly the door was swung opened and a terrified telegraphist rushed in. His news about radio communication failure added to onboard disturbance [b.16.2. 17.a]. The captain clearly understood the crew was waiting for his decision. They needed to do something to avoid a looming disaster.

16.2. 7. Watching terrified faces colored by a powerful green color, the captain noticed one more terrible thing. The color slowly turned darker [b.16.2. 17.b]. He glanced at the sky again and was unable to do anything more. The sky color darkened as the ship moved forward. Part of that darkness already reached the horizon, right on the ship's course, and connected to water [b.16.2. 17.c].

16.2. 8. The captain heard a cry next to him. He turned his attention to the pilot house. Mr. Jefferson gestured at the magnetic compasses and their failure. The captain grimace. He had never seen stranger indications from the compasses. Each of them gave different indications and slowly changed over time [b.16.2. 17.d].

16.2. 9. "Abandon ship!" The captain's command echoed from the walls of the pilot house. It was immediately translated to the crew crowded on the deck. But it was too late. Something changed again, and the men noticed an unusual lightness of their bodies. Each small step led to a large jump, and sailors rushing to the life boats jumped like leopards [b.16.2. 17.e].

16.2. 10. The telegraphist, trying to go to the deck from the pilot house, jumped down, but his effort had extra force due to the changes of gravity. He fell over part of the deck separating the pilot house from the nearest board and dropped into the dark green water.

16.2. 11. The captain and his first mate reached the deck with great care, watching each step. Walking brought an unusual sensation that was hard to imagine. Both men felt continued in the weight-lowered sensation. It was very difficult to control nearly weightless bodies.

16.2. 12. Cries of damnation reached the ears of both men. They turned their attention to the boat desk. A few men were watching a boat slowly moving away. Obviously, they tried to use the boat but were unable to control the almost weightless object. As a result the efforts led to losing the boat. Attempts of other groups of sailors had the same result [b.16.2. 17.f].

16.2. 13. Meanwhile the darkness became more and more dense. Surrounding objects were hardly visible through the dense mist. Human cries and splashing sounds of falling bodies in the water became [b.16.2. 17.g]. The captain tried to go further along the deck, but his effort was too much to keep balance. His body rushed through air and disappeared in the mist [b.16.2. 17.h]. The same thing happened to the first mate.

16.2. 14. A few seconds later both bodies had fallen into water to the huge surprise of their owners. According to their beliefs the deck of the cargo ship should be much higher. But both men found themselves in the salty ocean water [b.16.2. 17.i]. "Are you all right?" cried the captain to his fellow. "Yes, sir!" the answer came from the darkness. "Where is the ship?" asked the captain. "I have no idea!" answered the first mate [b.16.2. 17.j].

16.2. 15. Suddenly both men heard splashes of water around the stem [b.16.2. 17.k]. It was a familiar sound to them. Obviously, a ship was coming toward them. "Stand from under!" cried the captain, hearing splashes coming closer to him. A second later he saw through the condensed mist a huge stem and board coming right to him. It was the ship moving straight ahead. The men had no chance to survive under the steel beast.

16.2. 16. A few minutes later the ship left the dangerous area. Slowly anything onboard took its usual state, and nothing carried any sign of tragedy [b.16.2. 17.l]. The ship and onboard load was well, but there was nobody onboard. Because of the panic and the very unusual circumstances, the crew met their end. Only indications of onboard chronometers kept a sign of the disaster, but they were not noticed by the crew of the second ship that found the abandoned tanker a few hours later.

16.2. 17. Explanation of the [16.2. 2] event type.

a. It was a result of Rainbow Effect (more specifically LSRE). That effect disrupts radio communication before becoming visually noticeable. For more details, see [12.2; 12.3].
b. Gap-Hood changes the interaction level from full interaction to zero from the edge of Gap-Hood to Shield-Gap. As a result any phenomenon increases in the same way. For more details, see [9.4].
c. The closer the ship is to the Shield-Gap, the less interaction level that exists between the ship and CE-Space. As a result a visible darkness appears from some point of the ship's location due to a very low level of interaction between Gap-Hood at that position, and electromagnetic waves propagated from CE-Space. The same phenomenon blocks radar vision of any object located close to Shield-Gap (in Gap-Hood). For more details, see [12.4; 12.2].
d. The area of reduced interaction reduces the interaction with the magnetic field as well. As a result any magnetic device acts as a failed one. Changing indications of magnetic compasses are caused by changing level of interaction. The less level of interaction, the more the compasses interact with the magnetic

field produced by the electric equipment of the ship. For more details, see [12.5].

e. The area of reduced interaction reduces any type of interaction, including gravity. As a result the gravity field reduces continuously the closer the ship gets to the Shield-Gap. The only force of gravity interaction that keeps constant is the force of gravity interaction between the vehicle and its onboard objects. For example the lowest interaction between a man, with a body mass equal to 75 kg, and the ship, with mass equal to 50.000 tons, with a distance of 10 meters from the center of masses, has an interaction between himself and the ship equal to force $F = G \cdot (M_1 \cdot M_2) / R^2 = 6.67553 \times 10^{-11} \cdot (75 \cdot 50.000.000) / 10^2 = 6.67553 \times 10^{-11} \cdot 37,500,000 = 6.67553 \times 10^{-11} \cdot 3.75 \times 10^7 = 2.503 \times 10^{-5}$ [N]. "One newton is equal to a force of 100,000 dynes in the centimetre-gram-second (CGS) system, or a force of about 0.2248 pound in the foot-pound-second (English, or customary) system."[1] Hence, the force interaction between the man and the ship equals $2.503 \times 10^{-5} \cdot 0.2248 = 0.56 \times 10^{-5}$ pounds (instead of 75 kg, or approximately 225 pounds in CE-Space). In such a condition any muscle movement throws the body far away from the ship.

f. Reducing the gravity interaction leads to the weightlessness of any onboard object (including the boats, human bodies, etc.).

g. The area of reduced interaction does not blocks the movement of air molecules. As a result any noise is noticeable by human ears. Moreover, the sounds are able to go through the Z-Gate and can be noticed around the corresponding Shield-Gap. For more details, see [12.10; 12.11].

h. Located next to Z-Gate, any object has a unique ZT-Time that is necessary to go through Z-Gate. Generally two different objects with different sizes and speeds use Z-Gate at different ZT-Times. Therefore, a human body moving much faster than a ship has a lower ZT-Time and uses a Z-Gate faster than the slow ship. That leads to a relocation separated by time according to the IO's TKD (faster objects use Z-Gate in less ZT-Time). For more details, see [12.8].

i. That time humans successfully pass the Z-Gate. But the ship is still inside HE-Space. Because of its size and speed, it needs much more time to pass through Z-Gate (remember the calculation of ZT-Time). As a result the faster moving humans reached the corresponding Shield-Gap sooner than the ship. They had fallen into the water because of the absence of the ship at that time at the In-Gap point.

j. The ship was inside HE-Space at that time and was not noticeable from CE-Space.

k. After the necessary ZT-Time the ship appears at In-Gap.

l. Then the ship left the Gap-Hood's reduced interaction with CE-Space and establishes itself in real space again, and any onboard physical process behaves in its usual way.

16.2. 18. As we can see, the difference between the two types of Z-Processes described above focuses on crew location. If the crew is inside a closed space, they have a good

---

[1] "Newton." Encyclopaedia Britannica. Encyclopaedia Britannica 2008 Deluxe Edition. Chicago: Encyclopaedia Britannica, 2008.

chance to pass a Z-Gate without any problem. In that case the whole craft moves and relocates from an Out-Gap to the corresponding In-Gap as a single object (with humans inside).

16.2. 19. In cases of open vehicles (a ship with an open deck), crew members and a vehicle use the Z-Gate in different ways according to ZT-Time. As a result a ship and crew use Z-Gate that consequently leads to a separation of the crew from the ship. Such a separated crew has no chance to survive in the open ocean.

Conclusion

The long cruise mentioned at the beginning of this work has ended. I used my best efforts to explain everything related to Z-Theory as clearly as possible, in order to show these Z-Phenomena are not only findable and traceable around the world, but they are quite understandable by the human mind and are researchable regardless of any prevalent point of view.

However, most parts of the book are dedicated to the explanation of different phenomena related to Z-Theory, because without enough examples readers would not be able to understand most of the distant relationships between the surrounding world phenomena and the same physical process that I named Z-Process.

Appendix A

This appendix contains definitions of all the notions and their abbreviations that were used in this work. Each notion that relates to this or that Z-Theory application is marked by (ZTA). Other notions that have no marks belong to Z-Theory itself.

For each notion shows the number of its first entry (f.e.), where it was mentioned the first time.

A

ARI (Area of Reduced Interaction) (ZTA) is an area that has a lower level of interaction between physical units located in that area and outside of it. In Z-Theory ARI associates with Gap-Hood. f.e. [9.2]

AT (Altered Time) is the difference between indications of an IO's TKD that uses Z-Trajectory and an IB's TKD that keeps a location at CE-Space. f.e. [12.8. 19]

AT-Delay (Aphelion Time Delay) is the period of time that must be spent by a body moving in an elliptic trajectory to reach a point with the same value of S_g (Gravity Force Strength) passing the aphelion of an elliptical trajectory as the nearest point of upside. f.e. [11.1. 16]

'A-type' zone means Shied-Gap. f.e. [9.3. 3.g]

B

Back-Transposition is the opposite transposition of a moving system (object) by Z-Gate relative to the same ZTHP. f.e. [14.3. 7]

Buckshot Effect (ZTA) is the relocation of objects by Z-Process from the same point of space and the same point of time to "mirror" points (different points of space that relate to the elliptical trajectory of the celestial body) at a different time according to the IB's TKD and that is associated with ZTHP (In-Gaps). f.e. [15.12. 8]

B-Type Zone means Gap-Hood. f.e. [9.3. 3.h]

C

C-IB (Conservative Field of an Independent Body) is any combination of conservative fields of any independent body. Two conservative fields that are well-known today are the electric and gravity fields. f.e. [6.1. 3]

CE-Space (Clear Event Space) is one of two parts of space separated by an E-Shield. CE-Space is the part of whole space containing observable physical units. f.e. [9.1. 9]

CSI (Cross-Shield Interaction) is a type of interaction when physical units interact through a partly established E-Shield. In the case of a fully established E-Shield, no interaction is possible through it. f.e. [9.3. 3.a]

CSPS (Corresponding Star-Planet System) (ZTA) are any pair of solar systems (a star and its planet) with equal masses (the mass of the first planet equals the mass of second one, and the mass of first star equals the mass of second one). f.e. [15.4. 7]

CT (Cross-Trajectory) is part of the whole trajectory used a moving object (or system of objects) that leads through an E-Shield gap (crossing the E-Shield gap at the point of Shield Gap). f.e. [9.3. 10.c.]

D

D-radius (Radius of Detection) is the largest distance from which the spectator is able to have any sense of the happening event.

Dragon effect (ZTA) is the phenomenon of a living creature's appearance (by use of Z-Gate) at locations unreachable for them by RWT. f.e. [13.2. 37]

E

Edge-Experiment is an experiment that relates to areas that are not researched and described quite enough. f.e. [8.1. 13]

E-IB (Electric Field of an Independent Body). f.e. [6.1. 3]

Event is any process that occurs at some particular place (volume, area, etc.) and can be noticed by senses of someone or something. f.e. [2.1. 10]

E-shield (Event Shield) is an imagined place that separates the most distant place that is reachable for interaction between two or more physical units (particles, fields, etc.) from the area that is unreachable for interaction between them. f.e. [9.1. 7]

F

Fog Effect (ZTA) is the mistake of human vision in the observation of a reduced interaction area from the outside. That effect appears only when the whole visible electromagnetic spectrum is affected equally by the reduced interaction. f.e. [12.1. 10]

G

GB (Green Belt) is a specific layer in a galaxy disk that contains points with S_g that lie between the values of S_g at aphelion and perihelion of a star's galaxy orbit. f.e. [15.4. 2]

GH (Gap-Hood) is the zone where the level of interaction between CE-Space and HE-Space changes from zero to the maximum. f.e. [9.4. 4]

G-IB (Gravity field of an Independent Body). f.e. [6.1. 3]

GSE (Galaxy Spin effect) is the Spin Effect on a galaxy scale. f.e. [15.7. 13]

H

HB (Host Body) is the body that produces equal S_c (strength of conservative field) at any given point of SA-Outline and SB-Outline. HB can be a physical body or the equivalent replacement of any number of physical bodies producing the same conservative field. f.e. [7.2. 5]

HBES-Plane (Host Body Equal Strength plane) is the number of points that are equidistant from two exact points and lie between those points, while a perpendicular straight line connects those two points in the very middle of that line (that is the plane). f.e. [7.2. 8]

HE-Space (Hidden Event Space) is one of two parts of space separated by an E-Shield. HE-Space is the part of whole space containing undetectable physical units. f.e. [9.1. 9]

HS (High Shift) or **HSRE** (High end Spectrum Reduced Interaction) (ZTA) is the reducing of interaction between IO and the high end of the electromagnetic wave spectrum by their passage through a reduced interaction area. f.e. [12.2. 14]

HT-Delay (Half Time Delay) is the period of time that must be spent by a body moving in an elliptical trajectory to reach a point with the same value of S_g (Gravity Force Strength) passing the aphelion or perihelion of the elliptic trajectory. That period of time equals half of the period of a full revolution. f.e. [11.1. 16]

I

IA (Intrusion Area) or **BSIPDA** (Biological System Intrusion Possible Detection Area) (ZTA) is an area where the probability of a given biological system's detection exceeds zero. In the case of a Z-Gate used by a biological system, that system must be unable to reach any point of IA by RWT. f.e. [13.3. 18]

IB (Independent Bystander) is a thing (or person) that only observes an experiment and is never involved in the experiment (event). f.e. [2.1. 3].

In-Gap is a Shield-Gap where a moving object goes to (in) CE-Space. f.e. [9.4. 4]

INGA (In-Gap Area) is an area with the highest probable location of an In-Gap. f.e. [13.3. 6]

IO (Independent Observer) is a thing (or person) that observes an experiment and is involved in the experiment (event). f.e. [2.1. 3].

IS (Independent Spectator) is any independent bystander or independent observer. f.e. [2.1. 3]

L

LS (Low Shift) or **LSRE** (Low End Spectrum Reduced Interaction) (ZTA) is the reducing of interaction between an IO and the low end of the electromagnetic wave spectrum by their passage through a reduced interaction area. f.e. [12.2. 14]

M

MTT (Maximal Transposition Time) (ZTA) is the maximum value of time according to an IB's TKD that can be covered by Z-Trajectory from an Out-Gap located around the Earth (as a celestial body) in the past to an In-Gap located around the Earth at the present time. MTT relates to NMCE. f.e. [14.2. 45]

N

NDA (Negative Deviation Area) (ZTA) is any region with a lower value of S_g than the average level of the planet surface. f.e. [14.3. 2]

NMCE (Natural Mass Changing Effect) (ZTA) is the changing of a celestial body's mass by absorbing meteorites and interplanetary dust from space. f.e. [14.2. 2]

NTSE (Negative Time Shift Effect) (ZTA) is the allowed and increased probability by a negative gravity anomaly for an In-Gap's presence with a corresponding Shield-Gap located in the past. f.e. [14.2. 30]

NTT (Negative Time Transposition) is the relocation of an object (or system) to an earlier point of time relative to the IB's TKD caused by Z-Process. f.e. [12.14. 18]

O

Out-Gap is a Shield-Gap where a moving object goes from (out of) CE-Space. f.e. [9.4. 4]

P

PDA (Positive Deviation Area) (ZTA) is any region with a higher value of S_g than the average level of the planet surface. f.e. [14.3. 2]

PTSE (Positive Time Shift Effect) (ZTA) is the allowed and increased probability by a positive gravity anomaly for an In-Gap's presence with a corresponding Shield-Gap located in future. f.e. [14.2. 34]

PTD (Possible Detection Time) (ZTA) is the period of time that can be used by an IB for the observation of a live creature that passed through a Z-Gate. In other words PTD is the period of time when a creature exists at a new location (after passing an In-Gap). f.e. [14.1. 17]

PS (Passing System) is any number of physical objects passing through a Z-Gate simultaneously. These objects retain their usual interaction between themselves inside the Z-Gate (in Z-Trajectory). f.e. [13.1. 7]

PTT (Positive Time Transposition) is the relocation of the object (or system) to a later point of time relative to an IB's TKD caused by Z-Process. f.e. [12.14. 18]

PT-Delay (Perihelion Time Delay) is the period of time that must be spent by a body moving in an elliptic trajectory to reach a point with the same value of S_g (Gravity Force Strength),passing the perihelion of the elliptic trajectory as the nearest point of upside. f.e. [11.1. 16]

PU (Physical Unit) is any type of physical object or field.

R

Rainbow Effect (ZTA) is the changing color of light that is caused by its passage through an area of reduced interaction. f.e. [12.2. 11]

RB (Red Belt) (ZTA) is any part of the galaxy that can be described as a Green Belt but that lies closer to the galaxy center. f.e. [15.6. 4]

R-ZR (Red-Z-Ring) (ZTA) is a Z-Ring with a lower average radius corresponding to a particular Green Belt. f.e. [15.9. 66]

RWT (Real World Trajectory) is any trajectory where the position of a particle has only one exact point for any given moment of time. f.e. [4.1. 4]

S

SA-Outline is the number of outline points positioned closer to body A (applicable only in cases of two equal bodies A and B, or two different locations of the same body). f.e. [7.1. 35]

SB-Outline is the number of outline points positioned closer to body B (applicable only in cases of two equal bodies A and B, or two different locations of the same body). f.e. [7.1. 35]

SGF (Strength of Gravity Force) is the force characteristic that is equal to the value of the force produced by the mass of an object to the unit's mass positioned a given distance from the object. f.e. [3.2. 3]

SG also S-Gap **(Shield-Gap)** is the zone where full interaction between CE-Space and HE-Space exists. f.e. [9.4. 4]

SGY (Star Galaxy Year) (ZTA) or (Sun Galaxy Year in particular case) is the period of time that a star spends in one full revolution around the galaxy center. f.e. [14.3. 11]

SLVR (Star Light Velocity Rate) (ZTA) is a value that shows how many times the speed of a star is less than the speed of light. f.e. [15.1. 12]

S-Outline (Strength Outline) is the number of points with equal value of strength of a conservative field. The conservative field changes no energy of a particle by any motion of that particle between any number of such points by any trajectory. f.e. [7.1. 27]

Siren Effect (ZTA) is the phenomenon of audible sounds at some areas without their observable sources. The phenomenon is caused by a Z-Process in gases. f.e. [12.11. 6]

SSI (Side-Shield interaction) is the type of interaction between physical units on the same side of an E-Shield. In the case of a fully established E-Shield, only that type of interaction is possible. f.e. [9.3. 3.d]

Space-Time Transposition (STT) is the changing location of a moving system (object) by Z-Trajectory involving time according to a IB's TKD. Generally each transposition has this type. f.e. [14.2. 18]

Surge Effect (ZTA) is the process of the water level quickly changing in any given water pool that coincides with the same direction of thermal disturbance (increasing or decreasing water volume) caused by the presence of an S-Gap and a Z-Current inside the pool. f.e. [12.13. 7]

T

TKD (Time Keeping Device) is any type of device that is dedicated to measuring time intervals (chronometers, clocks, watches, etc.). f.e. [12.7. 8]

TT (Transposition time) is the period of time that is noticed by an IB as the time of "absence" of a system (object) that uses Z-Trajectory. f.e. [12.8. 19]

V

VB (Violet Belt) (ZTA) is any part of the galaxy that can be described as a Green Belt but that has a greater distance from the galaxy center. f.e. [15.6. 4]

V-ZR (Violet-Z-Ring) (ZTA) is the Z-Ring with the higher average radius corresponding with a particular Green Belt. f.e. [15.9. 66]

W

Window Effect (ZTA) is the IB's impossibility to have any evidence according to Z-Process and relating (derived) processes outside of appropriate time interval (observation window). f.e. [12.14. 17]

Z

ZA (Z-Appearance) (ZTA) is any phenomenon noticeable (observable) by an IB that relates to Z-Process or its consequences. f.e. [14.4. 24]

Z-Bridge (ZTA) is the combination of Violet-Z-Ring and Red-Z-Ring of two CSPS that can be used by a Z-Gate to establish interaction between them by means of the Z-Process. f.e. [15.10. 13]

Z-Current is the moving of liquid molecules from a given Shield-Gap to the corresponding one by evaporation and condensation. f.e. [12.12. 32]

Z-Event is any event that relates to Z-Process. f.e. [12.8. 6]

ZG (Z-Gate) is the combination of Z-Trajectory, two corresponding gaps (Out-Gap and In-Gap) linked by that trajectory, and two Gap-Hoods (surrounding both gaps). f.e. [13.1. 3]

ZLR (Zero Length Relocation) is the relocation by Z-Trajectory when the distance between first and last point of Z-Trajectory in a given frame of reference is equal to zero. f.e. [10.2. 13]

Z-Range (ZTA) is the combination of some CSPS inside the galaxy that relates to each other by the number of Z-Bridges. f.e. [15.10. 14]

ZP (Z-Process) is the process of passing through corresponding Shield-Gaps (Out-Gap and In-Gap) and using Z-Trajectory by any physical system (object). f.e. [11.2. 24]

Z-Phenomenon is any observable process related to Z-Process. f.e. [14.2. 63]

Z-Radius (ZTA) is the distance between In-Gap and the most distant point (in each particular direction) reachable for a particular type of creature that passes through a Z-Gate. f.e. [13.3. 15]

Z-Ring (ZTA) is the layer between two close stars' elliptical orbits with a different semimajor axis from the galaxy center, limited by the distance of perihelion of the orbit with the greater semimajor axis and the aphelion of the orbit with the lesser semimajor axis. f.e. [15.9. 46]

Z-Sector (ZTA) is the area on the planet surface with a gravity anomaly (positive or negative deviation of S_g from average level). Such an anomaly depends only on the inner planetary structure and is free from the influence of planetary motion and rotation. f.e. [14.2. 61]

ZT (Z-Trajectory) is the key notion of Z-Theory. Z-Trajectory is a special part of a moving object's trajectory that cannot be viewed or interacted with by the IB and always lies between two special points where the difference in the whole energy of moving object equals zero. As a result the whole energy of a system keeps constant before and after an object uses Z-Trajectory.

In other words Z-Trajectory is the image trajectory that connects two points of space with equal value of full strength of conservative fields by means of a special number of points that have no interaction with the surrounding conservative fields produced by independent bodies. f.e. [4.2. 6]

ZTT (Z-Transposition Time) is the minimal period of time that must be spent by a moving system according to the IO's TKD (moving object) that uses Z-Trajectory for full relocation from one head point of Z-Trajectory to the next one. f.e. [11.2. 22]

ZTHP (Z-Trajectory Head Points) are any points between which Z-Trajectory can exist. In CE-Space such points are always located at the S-Outline. f.e. [10.2. 17]

ZTTP (Z-Trajectory Tail Points) are any points of Z-Trajectory in HE-Space. Those points form Z-Trajectory itself in HE-Space. f.e. [10.2. 17]

ZTR (Zero Time Relocation) is the relocation by Z-Trajectory when the time delay between location of the object at first and last point of Z-Trajectory is equal to zero (result of mathematical calculation). In CE-Space such a relocation has very little time that almost equals zero. f.e. [10.2. 9]

Z-Window is period of time between first and last point of time of Z-Gap presence according to the IB's TKD. f.e. [14.4. 39]

Appendix B

This appendix contains a list of "strange" events that had occurred in different areas of the world. They were used in this book as real examples of Z-Process (transposition of objects). Such events are predicted by Z-Theory, and any direct observation supports that theory. All quotations have been taken from the book *Lo!* by Charles Fort.

List of Events

1. Lifeless Objects

B.1. 1. "London Times, April 26, 1821 – that the inhabitants of Truro, Cornwall, were amused, astonished, or alarmed, 'according to nerve and judgment,' by arrivals of stones, from an unfindable source, upon a house in Carlow Street." (p. 19)

B.1. 2. "London Times, Jan 13, 1843 – that according to the Courrier de l'Isere, two little girls, last of Desember, 1842, were picking leaves from the ground, near Clavaux (Livet), France, when they saw stones falling around them. The stones fell with uncanny slowness." (p. 20)

B.1. 3. "There is a story of this kind, in the New York Sun, June 22, 1884. June 16th – a farm near Trenton, N.J. – two young men, George and Albert Sanford, hoeing in a field – stones falling…. The next day stones fell again." (p. 21)

2. Living Creatures

B.2. 1. "There is an account, in the London Daily News, Sept. 5, 1922, of little toads, which for two days had been dropping from the sky, at Chalon-sur-Saone, France." (p. 10)

B.2. 2. "In the Redruth (Cornwall, England) Independent, Aug. 13, and following issues, 1886, correspondents tell of a shower of snails near Redruth." (p. 10)

B.2. 3. "Phyladelphia Public Ledger, Aug. 8, 1891 – a great shower of fishes, at Seymour, Ind. They were unknown fishes." (p. 10)

Appendix C

The Map

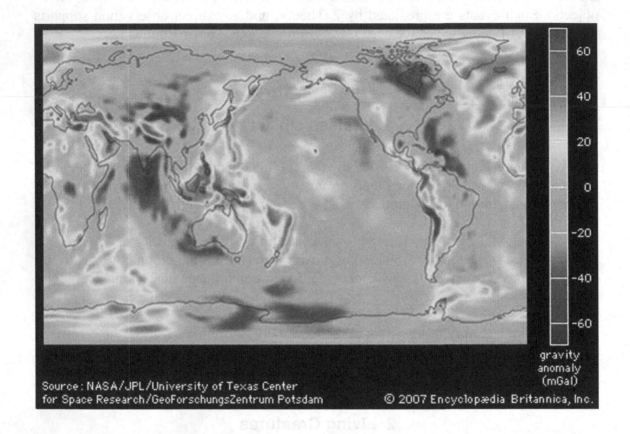

Source: NASA/JPL/University of Texas Center for Space Research/GeoForschungsZentrum Potsdam   © 2007 Encyclopædia Britannica, Inc.

"The variation in the gravitational field, given in milliGals (mGal), over the Earth's surface gives rise to an imaginary surface known as the geoid. The geoid expresses the height of an imaginary global ocean not subject to tides, currents, or winds. Such an ocean would vary by up to 200 metres (650 feet) in height because of regional variations in gravitation."[1]

---

[1] "Gravitation." Encyclopaedia Britannica. <u>Encyclopaedia Britannica 2008 Deluxe Edition</u>. Chicago: Encyclopaedia Britannica, 2008.

Appendix D

Sun Galaxy Year and Mass Extinctions

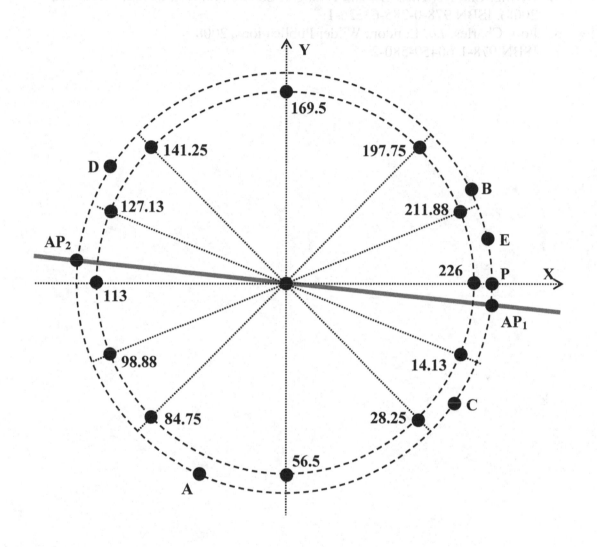

Bibliography

1. *Encyclopaedia Britannica, 2008 Deluxe Edition.* Chicago: Encyclopaedia Britannica, 2008 (electronic edition).
2. Berlitz, Charles. *The Bermuda Triangle*. London: Souvenir Press, (reprinted 2008). ISBN 978-0-285-63326-1.
3. Fort, Charles. *Lo!* London: Wilder Publications, 2008. ISBN 978-1-60459-580-2.